Raising and Keeping
Dairy Goats

A Practical Guide

Raising and Keeping
Dairy Goats

A Practical Guide

Katie Normet

FIREFLY BOOKS

A Firefly Book

Published by Firefly Books Ltd. 2017
Copyright © 2017 Firefly Books Ltd.
Text copyright © 2017 Katie Normet
Images copyright © 2017 Katie Normet, unless otherwise noted below.

All rights reserved. No part of this publication may be reproduced, stored in a retrieval system, or transmitted in any form or by any means, electronic, mechanical, photocopying, recording or otherwise, without the prior written permission of the Publisher.

First printing

Publisher Cataloging-in-Publication Data (U.S.)
Names: Normet, Katie, author.
Title: Raising and Keeping Dairy Goats : A Practical Guide /Katie Normet.
Description: Richmond Hill, Ontario, Canada : Firefly Books, 2017. | Includes index. | Summary: Detailed information on all aspects of goat farming for the small hobby farm.
Identifiers: ISBN 978-1-77085-979-1 (paperback)
Subjects: LCSH: Goat farming – Handbooks, manuals, etc. | Goats.
Classification: LCC SF383.N676 |DDC 636.39 – dc23

Library and Archives Canada Cataloguing in Publication
A CIP record for this title is available from Library and Archives Canada

Published in the United States by
Firefly Books (U.S.) Inc.
P.O. Box 1338, Ellicott Station
Buffalo, New York 14205

Published in Canada by
Firefly Books Ltd.
50 Staples Avenue, Unit 1
Richmond Hill, Ontario L4B 0A7

Cover and interior design: Firefly Books
Illustrations: George Walker

Printed in China

Canada
We acknowledge the financial support of the Government of Canada.

Photo Credits
Shutterstock: Antonia Giroux, 104; bafanny, 12; Buffy1982, 42; cosmopolit, 26; Edwin Butler, 33; Elena Lapshina, 150; Getz Images, 161; inspired by the light, 166; keellla, 77; Luca Santilli, 138; Pavelmw, 86; schubbel, 81; Tikta Alik, 54; urobot4, 112; Visun Khankasem, 49.

Cover Photos
(front) Thepphanom Leeprakhom/Shutterstock;
(back left) Buffy1982/Shutterstock; (back right) Katie Normet.

Contents

Introduction 7

CHAPTER 1
Getting Started 13

CHAPTER 2
Feeding Your Goats 27

CHAPTER 3
Breeding Your Goats 43

CHAPTER 4
Kidding 55

CHAPTER 5
Milking Your Goats 87

CHAPTER 6
Caring for Horns and Hooves 105

CHAPTER 7
Disease and Illness 113

CHAPTER 8
Cheese Making 139

CHAPTER 9
Soap Making 167

Resources 188

Index 189

Introduction

I RODE AND WORKED with horses for more than a decade before I met my first goats. The owners of the horse farm I rode at briefly added a herd of goats in an effort to make some much-needed income through sales of goat milk. One fateful day I walked into the horse barn and, instead of horses, there were goats—lots of goats.

I loved how alert these creatures were. They seemed to move almost as one, yet as individuals at the same time. Many had long, beautiful flowing ears that seemed to fly as the goats turned toward the sound of the grain scoop and strained to get as close as they could to the feed. Their body language seemed to cry out, "Me, please, me, please!" They seemed to be an equal-opportunity species, and I was curious and eager to learn more.

At the time, I was in third year at the University of Guelph, in Ontario, where I studied animal science and had courses in food processing, food microbiology and food chemistry. Most of the food science I was learning was about milk. When I signed up for food science courses, I had no idea I would be learning so much about dairy. At this point in my life, I had not even considered the possibility of milking goats and making cheese. I took the food science courses because I was attracted to the science, but I had no clue what I would actually do with my acquired knowledge. During my years at Guelph, my fellow classmates brought one aspect of farming to my attention: There is no expensive quota needed to sell goat milk.

In Canada, if a farmer wants to produce cow milk, he or she must purchase one quota per dairy cow. Originally, quotas were free; today, a single quota (which allows the owner to have one producing dairy cow) may cost tens of thousands of dollars, depending on the province in which you're farming. Breaking into this system is an expensive undertaking for new farmers, and learning that dairy-goat farming did not come with such a huge price tag was good news to someone with a growing interest in farming for a living.

The pieces of my life were falling into place. My new love for goats, my new passion for dairy science, my discovery of the lack of need for quota, my desire to not work for someone else, my lifelong desire to live on a farm ... it all came together and clearly pointed out to me that I should milk goats and make cheese.

In 1999, with my parents, my partner and I purchased an abandoned 95-acre farm just outside the village of Arthur in southwestern Ontario. The farmhouse had not been lived in or cared for in more than 50 years. There was no electricity, which meant no running water at the house. The derelict barn had been used for grain storage for at least a couple of decades; its rat population was well fed. The existing concrete pens in the barn were several feet deep with ancient manure, while one large section of the hayloft had fallen to ground level.

During the first years on the farm, I stayed home with our toddler son while my husband worked off the farm. Goat farming was our long-term plan. Meanwhile, we cleaned up the property, built fences and cleared the barn to keep cattle and veal calves. We raised hens for eggs and chickens for meat, and I honed my farm management skills.

Creating a suitable home for goats was a slow and steady process, but by March 2001, we purchased two very affordable Alpine does from a hobby farmer in the area.

In hindsight, I realize these two goats were really more a science experiment than anything else. Our does were healthy-looking animals and of good weight, and the fact that they were already bred was also appealing. After we'd loaded the goats safely into the truck, the farmer told us their breeding dates. On the drive home, I calculated the kidding date, admittedly a little late in the game. "They're due now!" I exclaimed.

We arrived home with the goats early that evening, and when I checked on the does at 11 p.m., one of them had four kids on the ground! Around midnight, the second doe delivered a second set of quads. It was quite an introduction to the goat world. The two does, affectionally called Auntie

Author's daughter Clara and her pet goat Pippa.

and Grey, were fantastic milkers. They raised all those kids—six bucks and two does—and still had milk left for us. I was happy with the eight "instant goats," though not quite as happy that there weren't more does offering future milk prospects.

By September of that year, we'd set up our farm to milk goats commercially. The province granted us our Grade A milking license, and we could legally start shipping milk. We purchased a single herd of 20 crossbred goats, and the milk broker's milk truck arrived once a week to pick up the milk we produced, delivering it to a large commercial cheese plant nearby. Shortly after we started selling our goat milk, however, we grew anxious about the stability of Ontario's goat-milk industry. I came up with a solution: goat-milk soap. I knew how to milk goats, and I understood the science of soap making. I signed up for a one-year business course offered by the Canadian government to learn more about how to run a retail business. Our own family was growing. With a four-year-old son and a one-year-old daughter, I was grateful we had parents who were willing to babysit.

In 2003, River's Edge Soap Company was born. Over the next years, my mother and I spent countless weekends traveling around southwestern Ontario as we sold of bars of soap, goat-milk moisturizing creams and related products. During our travels, we were asked the same question over and over: "Do you sell goat cheese?" We didn't, but I knew I wanted to. It was time to fulfill our dream and build our own milk-processing plant.

We renovated a room just off our milk house. The space had to meet the strict regulations needed to sell goat dairy products to the public. We purchased cheese-making equipment, which included a custom-made pasteurizer designed by my civil engineer father, a very basic cheese vat—a stainless steel soup pot purchased from a discount kitchen supply store—a scale, an assortment of stainless steel utensils, a batch of cheesecloths, temperature recording devices and a thermometer. On July 1, 2005, we were granted our provincial license to run a goat cheese-processing plant. River's Edge Soap Company evolved into River's Edge Goat Dairy. We could now legally produce and sell our goat-milk products in Ontario.

Making cheese in our new processing facility was certainly different from making it in the family kitchen, as any restaurant chef reducing a recipe for a hundred customers down to a family-size meal will attest. After some trial and error, I perfected the technique for making pasteurized milk, yogurt, chèvre and feta. But I knew I had stretched my self-learning to its natural limits. It was time to go back to school.

In 2008, I made my first visit to the Vermont Institute for Artisan Cheese. Since then, I've returned to that inspiring state a dozen times to take courses and continue my research. The expertise I gained in Vermont has resulted in huge benefits for the River's Edge Goat Dairy in both the making and the marketing of our cheese. During the summers of 2011 and 2012, I perfected recipes for two mold-ripened and three semi-hard cheeses. That was a bit of a challenge since my new partner and I added three children to the family, bringing the tally to three boys and two girls.

In 2015, we milked 75 goats and produced more than 50,000 liters of milk, which we turned into our delicious goat-milk products. Today, we sell our products directly from the farm and at farmers' markets in Guelph, Kitchener and Orangeville, Ontario. Each week, we sell approximately 200 liters of fluid milk, 40 liters of yogurt, 20 kilograms of chèvre, 20 kilograms of feta and an assortment of Pippa, Camembert, hard cheese and blue cheese. River's Edge also sells goat milk soaps and moisturizing creams as well as goat-meat products and our unique goat butter tarts.

We've come a long way in the past two decades. Owning and working with goats is a joyful mix of hard work and reward, an extremely satisfying and family-friendly project. The key to getting the most enjoyment and production from your goats is to learn as much as possible about them before introducing them to your lives. Once goats become a part of your family, they will teach you even more.

Before you make the commitment to purchase and add goats to your life, realize a few things.

- Goats need a minimum of attention every morning and every night. Is someone available to "do chores"? That means making sure the goats have adequate feed, water and bedding and the right environment. You'll be on call mornings before work or school, and later after a long day.
- The longest goats should be without someone to check in on them is about 12 hours. If you plan on being away longer, a goat sitter is necessary. It's difficult to go on last-minute vacations when you have goats at home.
- There can be no procrastinating when it comes to milking or feeding. The work must be done on time. Goats don't do well when made to wait.
- You must be prepared to miss, or be late for, a meal, a night's sleep, an important business meeting, a wedding, a soccer game or other life events in order to care for your goats.
- You must be physically capable of working with goats, lifting pails of milk and bales of hay.
- You must live in the right location to own goats. Many municipalities are now allowing goats on non-agricultural land. Check with your local government to be sure you're allowed to house a goat on your property.
- You also must have adequate space and facilities for your goats to live comfortably.

The decision to get goats should be made by the family. It's helpful and will cause less tension if everyone in your family is on board.

In these pages, I offer a general view of goat rearing for those who have little or no knowledge of goats and a review for those who already have some goat experience. I'll share what I've learned about raising and housing goats, feeding them and keeping them healthy, as well as my techniques and recipes for cheese making and soap making. Whether you're simply keen to provide some goat milk and cheese for your own family or interested in launching your own operation, I trust you'll find helpful advice here.

Katie Normet
River's Edge Goat Dairy

CHAPTER 1

Getting Started

WHILE THERE ARE more than 200 goat species in the world, only a handful of breeds are raised by goat farmers in North America. Farmers select goat breeds according to the quality and quantity of a breed's milk as well as its butterfat content. While descriptions of these qualities exist for each breed, always keep in mind that the taste of milk is subjective and that every goat is an individual, so both production numbers and milk flavor will vary.

Acquiring Goats

It takes a great deal of experience to raise healthy, happy goats through the life stages of breeding, pregnancy, kidding, kid care and milking management. If you have never raised goats before but are ready to take the plunge, I have two suggestions. The first is that you start slow, with two goats. The second is that you purchase young kid goats rather than adults. Take baby steps until you've achieved a level of comfort before taking on all the challenges of a mature goat. By raising these kids yourself, you'll gain invaluable experience. Caring for a pregnant doe or a doe already in milk when you have little or no experience could prove a disastrous and expensive undertaking.

When it comes to your search for goats, your biggest asset is patience. There are many routes to purchasing goats, but they all take time. Talk with your neighbors, check in with your vet, and keep an eye out for classified ads and ads posted at your vet's office or general store. Take weekend drives around the countryside and drop in at county fairs to see the goat shows. There are also specialty magazines and websites devoted to goats and goat breeds.

Questions to Ask Before Buying Your Goat

What type of goat should I buy? Over a long history with goats, humans have succeeded in creating breeds that satisfy our specialized needs. Today, there are three central goat types: dairy goats, which are bred to produce milk; fiber-producing goats, which are bred to produce mohair, cashmere or angora for shearing; and meat goats, which are bred to produce meat (also known as chevon).

Although a female goat that is bred for meat, such as a South African Boer or a Kiko, also produces milk, it usually produces no more than a liter/quart a day for no more than three or four months—basically enough to feed her own kids. A dairy goat, in contrast, produces one to six or more quarts/liters a day for a minimum of 10 months and up to two years.

Dairy goats can also be used for meat, but there is not much meat on a goat bred to produce milk. A dairy goat puts on very little subcutaneous fat (fat that sits just under the skin), while a meat goat is quite muscular and becomes plump eating the same quantity of feed as a dairy goat. A goat bred for fiber also produces milk but, as with the meat goats, only enough to feed her kids. A fiber goat also produces meat but, as with a dairy goat, it doesn't put on quite as much muscle as does a goat bred for meat.

There are also dual-purpose breeds. A heavy, meat-type Nubian won't produce as much milk as a dairy goat, but she produces an acceptable quantity of rich milk for the better part of a year as well as muscular kids. There are also some dual-purpose fiber-meat goats, such as the Spanish goat.

If your goal is to provide a family of four with milk, a single dairy goat will probably suit you fine. Depending on the breed, genetics, stage of lactation and management, one dairy goat can produce between one and six quarts/liters of milk a day. If your family requires 2.5 liters (quarts) daily, one goat should be enough. If milking a goat every day seems too demanding and you want just a little milk to

Lively, active, jumping kids using a patient doe as a climbing place.

experiment with soap making, then a meat or fiber-type goat should suit your needs.

Nevertheless, I recommend that you purchase two female goats rather than one. A single goat soon becomes incredibly lonely and depends on its human caretaker for attention and companionship. There isn't a barn or a fence in the world that can contain a goat desperate for company. It's also true that two goats may do the same, but it's far less likely.

With two happy female goats, it is more likely that at least one will be producing milk at any given time. Similarly, the chance of both failing to kid is unlikely.

Do I need to purchase a registered or purebred goat?

If you're not going to show your goats in competitions, do not spend extra money on a registered animal. Purebred, or registered, goats don't produce any better than unregistered or crossbred goats—sometimes called "commercial" or "grade" animals.

Unless you want to show your animals, the important factor in purchasing a goat is how much milk she produces.

I love crossbred animals. Genetic variation makes an animal less susceptible to breed problems. Although purebred goats don't have the extreme difficulties that some breeds of dogs have, such as hip dysplasia or susceptibility to certain cancers, crossbred goats tend to be a little hardier and stronger than purebreds.

Does the size of the goat matter?

Personally, I like goats that are on the smaller side. When we first started milking goats, dairy goat producers would visit the farm and laugh at our "small goats." One of these same farmers once asked us to milk a few of his goats for him over the winter. He owned big purebred Saanen goats. Those goats easily ate twice as much feed as our small ones, yet they produced the same amount of milk—and sometimes less.

Should I purchase a goat that's already milking?

You can purchase a goat that is already milking, but I strongly discourage people from doing that. More often than not, you end up with someone else's problem. Generally speaking, nobody likes to sell good animals—they would rather keep them for themselves because good animals are hard to come by.

Keep in mind, too, that goats do not like a change in routine. If you remove a mature milking goat from her home, herd and familiar feed, she will be stressed during the transition. Stressed goats do not produce well, and goats can drop significantly in milk production or stop milking altogether when moved to a new home. We once had a goat that stopped milking because we renovated our milking parlor. She simply couldn't handle the change. She stood in her old place by the door—though her "place" was no longer there—shaking the whole time. After a few days, she quit producing milk.

When purchasing a mature goat, you're stuck with what you get. Buying a kid that will grow up to be exactly what you're looking for is a much easier and more affordable strategy.

How old should a kid be when purchased?

You can purchase a kid at different stages. She could be a few days old, a few weeks old but still on milk, or several weeks old and weaned from milk and eating solid feed. The age of the kid is not as important as understanding what to do with her once you have her.

Don't worry so much about the appearance of the barnyard at the farm you're visiting. Raising goats takes a lot of time, energy and effort, and most goat farmers also work off the farm. Chores like grass cutting or maintaining a nice landscape often come after the care of the animals, as they should. Glance past the barn and barnyard and look at the goats and their immediate surroundings instead.

What is the air quality like in the barn? Goat barns have a particular odor, but it shouldn't be offensive. If you're uncomfortable breathing in a barn, the goats are uncomfortable as well. Goats are highly susceptible to pneumonia and lung damage from poor air quality and may have a permanent cough from being in such an environment for extended periods.

Look at the bedding, if there is any. Is what's there plentiful and dry? Wet bedding is more likely to house large amounts of harmful bacteria, molds and fungi than dry bedding.

Look at the feed in the mangers. Is there an abundance of feed, a small amount or none at all? Look at the quality of feed. Is the hay green and dust-free? Is there access to pasture?

Look at the goats themselves. Do they have nice, sleek hair coats? If they are fiber goats, does the hair look healthy, with an almost wet look to it? Patchy, dry-looking, rough hair is a sign of poor health in a goat.

What is the body language of the goats? Happy goats lounge peacefully, eat with enthusiasm, play together and are curious about the new visitor in their barn. If goats appear listless and are not doing anything, they are in poor health.

Look at the herd as an entity. Almost every farm has one or two goats that aren't 100 percent healthy at any given time, but a goat's health can turn very quickly, going from happy and energetic in the morning to near death in the afternoon. If you feel uneasy about the goat's health, pay attention to your gut; thank the farmer for his or her time, and walk away empty-handed. You can always visit other farms to get a better idea of what you're looking for and return later if you think you may have misjudged the situation.

Kids can be reared in one of two ways. The caregiver can keep them in a separate pen and bottle-feed them, or they can stay with their mother and self-feed. Because disease can be passed from doe to kids through the mother's milk, a farmer may remove the kids from the doe at birth and bottle-feed them pasteurized colostrum or cow's colostrum. Afterward, the kids are bottle-fed pasteurized milk, cow milk or milk replacer. If a farmer is practicing disease prevention from doe to kid, the price of the kids will be higher because it is a much more labor-intensive process.

If the kids are sleeping when you visit the farm, have the farmer wake them up. Don't go in the pen, however, unless the farmer invites you. A healthy, growing kid stretches upon standing up. A wagging tail is another sign of a healthy, happy kid. Kids may not always get up and run, jump, bounce and play, but if they do, you know you're looking at a group of healthy animals. They should have fluffy, clean hair coats. A dirty, yellowish stained back end is a sign of diarrhea. Kids should appear to be a good weight. If you can see defined bone structure, then kids have not been getting the nutrition they need. Bright eyes and perky ears are also signs of good health. This is one case where you do not want to pick the poor kid lying off on her own.

How old are the kids?

There are clues, but it's still good to verify the age with the farmer. If the umbilical cord is still present but dry, the kids are less than one week old. If the umbilical cord is still wet or damp, they are less than 24 hours old. Kids are disbudded at seven to 10 days of age. If that has already happened, you may see the burn marks on the kids' heads. If the horn buds are covered with hair, the kids are probably three to four weeks old. If kids have not been disbudded, small horns will start to grow at about two weeks of age. If they have horns that are a couple of inches long, the kids are most likely six to eight weeks old.

Who are the parents of these kids?

Ask if the buck is present and if you can see him. Ask which does gave birth to these kids. The farmer may know which kid came from which doe. If it is a larger farm with many kid goats, the farmer will know what group of does the kids came from because they'll be the does that recently came into milk.

What is the kids' current feed?

Goats don't thrive physically when their feed is suddenly changed (or when their routines are changed), so, when starting out, it's important to keep the kid on a familiar feed. In addition, this information will help you determine whether the kids have had a good start in life. If they are less than four to six weeks old, they should still be getting milk in some form. If you're prepared to bottle-feed, getting a kid that's still on milk is the perfect bonding opportunity for you and your family. Unless the kids are very young, they should be on a grain ration of some form—ideally a goat starter ration, though one formulated for cows will do as well. Whatever the ration, ask to take a small amount with you, and ask as well for the name of the feed and its manufacturer. You may be able to find the same feed but, if you can't, help the kid with the transition by blending the former grain with the new goat starter ration.

What is the kid's gender?

Before you choose your kid or kids, know whether you're looking for a dairy goat or a meat goat. If you want a milker, you need a female goat, which means you need to know whether the kid you're looking at is male or female. It can be tough to identify a kid's gender, especially when it is very young. Both bucks and does have nipples, or teats. Testicles may not be descended, or it may be difficult to feel them. Conversely, a doe kid's udder is sometimes slightly swollen because of circulating hormones, and the udder may feel like testicles. I've seen people mistake an umbilical cord for a penis. The best way to sex a kid goat is to look under the tail. A buck kid has nothing but an anus, which is situated right up underneath the tail head. A doe kid has an anus and a vagina. The vagina looks like a little pink "v" under the anus. If that's what you see, you're certain to have a female.

How much colostrum did the kids receive, and what was the source?

In the first 12 to 24 hours of life, a newborn kid requires only colostrum, the first milk that is produced by the maternal doe. Colostrum is rich in fats, proteins, milk sugar and the antibodies necessary to jump start a newborn kid's immune system. If the farmer is feeding the kids pasteurized colostrum, cow's colostrum or dried colostrum, be prepared to pay a premium for the kids. This disease-prevention measure is well worth the extra money. If you purchase a kid that is less than 24 hours old, be sure you take an adequate supply of colostrum with you just in case the farmer does not have a supply. Dried colostrum is an alternative and should be available at your local livestock feed store.

What type of milk are the kids drinking?

If yes, ask which brand and type, and make a note. Or is it goat milk, pasteurized goat milk or cow milk? If the kids are drinking raw goat milk, the possibility of transferring disease from doe to kid is much greater. Kids separated from does at birth and fed an alternative milk source or pasteurized milk and colostrum have the lowest risk of developing an incurable disease.

Have the kids received vitamin E, selenium or vitamin D shots?

Common nutritional deficiencies can cause problems with muscles and joints in kids. While these shots are not necessary, especially if the does are on a sound nutritional plan, they are a good preventive measure. And if your kid goat does get sick at some point, your veterinarian will probably ask whether the kid has received these shots.

Are the kids vaccinated?

Kids should be vaccinated for CDT to protect them against infections from *Clostridium perfringens* type C and type D and tetanus at four to six weeks of age and receive the booster three to four weeks after that. The risk of overeating disease is diminished if this vaccine is given.

Breeds of Goats

There are eight breeds of dairy goats common to the United States and Canada: Alpine, LaMancha, Nigerian Dwarf, Nubian, Oberhasli, Saanen, Sable and Toggenburg, though the popularity of a rare breed called the Guernsey is slowly increasing in North America. The breeds described below are standard-sized except for the Nigerian Dwarf. There is another category of goat that is described as "grade" or "experimental." These goats are typically crossbreeds; that is, the offspring of parents of two different breeds. They may also be the offspring of parents that represent more than two breeds. A "recorded grade" is a goat whose pedigree is recorded with the American Dairy Goat Association (ADGA) but is not registered as a purebred.

Alpine

With a reputation as a good milking breed, the medium- to large-sized Alpine dairy goat appears in a range of colors and patterns. It is friendly, curious, hardy, adaptive and sometimes strong-willed. It has a straight nose and erect ears, and both sexes have horns. The does should be 76 cm (30 inches) tall at the withers with a weight of 61 kg (135 pounds) or more; a buck should be at least 81 cm (32 inches) tall and weigh 77 kg (170 pounds) or more. At up to 3.5 percent, the butterfat content of the Alpine is comparatively low.

Alpine Crossbreed

LaMacha/Cross

Its production while lactating is three or four liters per day and, over nine to 10 months, is roughly 1,000 kg (2,200 pounds) of milk.

LaMancha

The quiet, friendly and curious medium-sized LaMancha dairy goat comes in all colors and is the only goat breed developed in the United States. Its hair is short and glossy. It is recognizable for its slightly dished profile and its tiny or non-existent external ears. Gopher ears can be up to 1 inch long; elf ears can be 2 inches long. A doe can have either; a buck can't be registered unless it has gopher ears. The doe is 71 cm (28 inches) at the withers and weighs 59 kg (130 pounds) or more; the buck is 76 cm (30 inches) at the withers and weighs 70 kg (155 pounds) or more. The does produce less than 1,000 kg (2,200 pounds) of milk with an average butterfat content of 3.8 percent.

Nigerian Dwarf

A miniature goat breed that originated in West Africa, the gentle, trainable Nigerian Dwarf comes in a range of colors, including white, cream, red, black and patterned. It can have a slightly dish-like face, and most have horns, though these goats are usually disbudded at a young age. For its size, the Nigerian Dwarf doe produces a high volume of milk, averaging 1.1 kg (2.5 pounds) a day with a high butterfat content. The doe's maximum height is 57 cm (22.5 inches) while the buck is roughly 60 cm (23.5 inches). This goat's average production of 324 kg (715 pounds) with a high butterfat content of 6.5 percent makes it an excellent choice for both cheese and soap making.

Nubian

The ancestry of the Nubian goat goes back to Asia, Africa and Europe. With long pendulous ears and a Roman nose, the Nubian is vocal with a convex facial profile and comes in a range of colors and patterns. Known for its high-quality milk with its high butterfat content, averaging 4.7 percent, the Nubian is considered assertive and fearless. The doe is at least 76 cm (30 inches) tall at the withers and weighs about 61 kg (135 pounds); the buck is 81 cm (32 inches) tall and weighs about 73 kg (160 pounds).

Oberhasli

Originally called Swiss Alpine in the United States, the medium-sized Oberhasli dairy goat has a uniform red bay coat that ranges in color densities, though the does may be solid black. It has two black stripes on its face from above each eye to its black muzzle, with an almost all-black forehead. It has a deep jaw, wide muzzle and

Nubian/Saanen Cross

forehead, and prominent eyes, with a face that is either straight or dished. The doe is about 71 cm (28 inches) at the withers and weighs about 54 kg (120 pounds); the buck is at least 76 cm (30 inches) and weighs at least 68 kg (150 pounds). In short, this is a quiet, dignified breed with an attractive appearance.

Saanen

Originally from Switzerland, the Saanen is the most popular Canadian dairy breed and now the second most popular dairy goat in the United States, after the Nubian. Solid white or light cream in color with a straight or dished face, this breed is one of the largest of all breeds, reaching 61 kg (135 pounds) and 76 cm (30 inches) at the withers. Sometimes referred to as the Holstein of the dairy goat world, it is an excellent milk producer, though at around 3.2 percent, its milk is fairly low in butterfat. In an average

Horns and Udders

Both male and female goats have horns. Every male goat has a beard, while the female goat may or may not have one. If a female does have a beard, it is much smaller than that of a male goat. Goats can also have wattles, seemingly useless bits of skin that hang from the goat's neck. Udder conformation in a goat is also important—it's desirable to have female goats with a well-attached udder and two teats.

lactation period, a doe produces some 1,135 kg (2,500 pounds) of milk, which accounts for the Saanen's popularity with dairies.

Sable
The relatively rare Sable is basically a Saanen that appears in any color other than white or cream. It shares all other characteristics with the Saanen.

Toggenburg
Named after the Toggenburg valley in Switzerland, the Togg is the oldest registered dairy goat breed. It is light to dark chocolate brown in color, with distinct colorings that include white ears with a dark spot in the middle, two white stripes down its face and white on its legs from about the knees down. It is medium-sized, with the doe standing 71 cm (28 inches) at the wither and weighing 54 kg (120 pounds). The doe produces an average 1,000 kg (2,200 pounds) of milk with 3.2 percent butterfat.

Crossbred goats
There is no one breed that ticks all the boxes on our list of wants in a goat. For this reason, we have crossed different breeds to create the perfect goats for us that thrive on pasture and produce high-quality milk. I like the Nubian or LaMancha crossed with the Saanen or Alpine. Our herd was established in 2002, so we have a strong foundation. Most of our milking goats have genetics from three or four different breeds. We then rotate the breed of buck that will be the father of our next generation of goats. We have fun picking our breeding bucks. We examine the health and management of the buck's herd before making our purchase, looking for traits that blend well with our own goats. While it is important to choose a good-quality breeding buck, good management will, ultimately, provide far greater value to your herd than even the best genetics. No matter what breed you choose, you'll make your goats your own.

Other Considerations

Housing
The type of goat will dictate the kind of housing you need to provide. Because dairy goats lack a thick hair coat and excess fat, they are not good candidates for a cold winter outdoors.

A cozy barn is a must for a dairy-type goat. The dual purpose dairy/meat goats and fiber goats with their long hair can tolerate being outside in the cold winter months but will still need shelter—especially from rain, strong winds and extreme cold.

Goats can tolerate cold temperatures, but not drafts and dampness combined with the cold. Basically, goats thrive in a clean, dry, draft-free but well-ventilated environment. Because goats are social animals, they do well in a common pen, where they're free to move about and meet and greet one another.

At our farm, we keep new kids in small pens built from new plywood, with about 10 kids to a pen. They're bedded on shavings, and the bedding is cleaned and topped up every day. Straw is also a bedding option. The pens are filled as the kids are born, so each pen contains kids that are roughly the same age. We use heat mats and heat lamps to provide warmth, as well as a plywood "roof" during the coldest weather.

We have a seasonal goat herd, meaning all our goats are bred and then kid at the same time only once a year. For us, kidding occurs during the months of March and April. In March 2015, we had some nights that dropped to −28°C (−18°F). We had to get our kids in boxes on heat mats with heat lamps. The lesson to be learned is that, as an animal owner, you must adapt to a situation. Weather in our region doesn't usually get so cold in March, but we can't change it and must adapt. We worked long hours to accommodate our goats so that they'd thrive.

When the kids are weaned and safely on hay and grain, at about six weeks old, we move them into group housing in larger pens, though bucks and does are always kept separate. When the nice weather arrives, they have access to pasture as well.

A number of housing options are available today for would-be goat farmers, from small sheds to hoop housing framed with tubular steel and enclosed in plastic canvas. Many people simply adapt a building that already exists. Some farmers may cut a smaller door in their barn door to allow goats to come and go into a fenced yard.

Comfort for the human caregivers is also an important consideration. Good lighting, accessible storage areas for feed and bedding, access to tools and medical equipment, electricity and running water, passageways and doorways that are large enough to accommodate a wheelbarrow—these things go a long way to making raising goats a pleasant and practical activity.

Fencing

Goats famously climb and seem magically able to get up and over many fences and can occasionally unlatch a gate. Six-foot chain link fencing is expensive, but it's secure enough

Young goats love to be up high! Pens with climbable features are a favorite with goats.

to baffle most goats. Other options include stock fencing, field fencing and electric fencing. Your decision will depend on your property and your budget.

Slow and steady wins the race

Thinking you'd like to start with a bigger herd of goats? Once again, I caution you to start slow. It's a big job feeding and looking after the goats, milking goats, keeping equipment clean and organized, and making your dairy products and managing them. Once you have a grasp of the tasks involved, milking goats and processing the milk will be second nature and possible efficiencies will become apparent.

So start small. Invest as little money and time as possible creating the housing for your first two goats. You'll have more expenses than you realize in your first year. As you learn what's needed, you can plan and design a better system that works for you and your goats. Keep in mind that your females should give birth at least once every 12 to 18 months, depending on when you want to breed and kid out your goats. Goats give birth to one to five kids, so a well-managed herd will naturally grow on its own more quickly than you think.

CHAPTER 2

Feeding Your Goats

BECOMING AN EXPERT in dairy goat nutrition can be a lifelong undertaking. While I can't tell you everything there is to know about proper goat nutrition and diet in these pages, I can share what you need to know to raise happy, healthy goats.

The goal of feeding is to have your animals at an optimum level of nutrition during all stages of production and all seasons of the year.

The feeding guidelines I describe are based on goats that are being raised in a traditional farm setting with pastures and fields for crops. Goats are browsers rather than grazers. In fact, eating a mix of shrubs, leaves, bark, weeds and grasses is what a goat is designed to do. As they browse, goats intuitively pick the most nutritious parts of these plants—typically the newest growth. They also naturally satisfy their need for the fiber in rumination with the more mature stalks of grass and twigs.

If you're lucky enough to have land that's suited for browsing, you're a fortunate goat owner. Knowing your land is an important part of your dietary management. If you milk according to your goat's natural breeding cycle, she will kid in the spring and come into peak lactation at the same time as your browsing land is at peak nutritional value. As the plants become more mature in late summer and fall and decline in nutritional value, milk production naturally drops. By winter, when the plants are dormant, your goat should be dry as well. It's a beautiful natural breeding-and-lactation cycle that's dictated by the sun.

As with any livestock species, feeding goats is a marriage of art and science. It's possible to feed your goat entirely by the book, using forage and grain analysis, body weight and weight of feedstuffs. But if you do, it's almost certain that, at some point, your goat will suffer.

For the best results, the facts and numbers of science need to be blended with the art of careful observation. Hay quality, pasture health, a goat's body condition score (BCS) and behavior and the stage of lactation cannot be easily evaluated and quantified. These assessments require a skilled eye—one that improves with every passing season. Hopefully, gut instinct will also eventually kick in. Remember, too, that the specific cycles of pasture and browse land and those of your goat occur only once a year. In effect, if something isn't working, you must wait 365 days to take a different approach.

One of the keys to becoming a better goat manager rests with the observations and thoughts you record in your goat journal. These can serve as a road map to maintaining the productivity and health of your goats in the upcoming year. For example, if you've noted that growth in your

Elevated grain and hay feeders with a step fulfill the goat's natural desire to reach for their food while keeping their feet out of their feed.

pasture or browse land slowed significantly when there was less than one inch of rain in a two-week stretch, you can prepare for this scenario in the future by having supplement feed on standby and thus preventing a drop in milk production. If you wait until your goat drops in milk production before you make a feed change, chances are your goat will not bounce back to pre-drought production levels.

Large commercial goat operations, whether for meat, milk or fiber, regularly test their feedstuffs for nutritional content. The results allow the farmer or nutritionist to formulate a feed ration (the total combination of carbohydrates, fats, proteins, minerals and vitamins—if lacking—and forages) for the goats. This practice makes economic sense if you're feeding a large herd of goats that represents your primary source of income. For the backyard goat farmer, however, learning what good-quality grains and hay look like is a far more valuable tool than relying on a lab to tell you exactly what is in a grain or hay crop.

So, how can you identify good-quality grains? Here is a few pointers:

Good-quality hay is hay that is free from mold, dust, mildew and off-odors and has a nice green color. It has not been rained on between cutting and baling. The greener the hay, the better, but note that the presence of weeds can make hay appear greener. Be sure the grasses and alfalfa are green, rather than the weeds. The younger the grass is at cutting, the higher the hay's nutritional value. As grasses mature, digestible fiber decreases while indigestible fiber increases. Protein levels also decrease as grasses mature.

When comparing hay of similar maturity, remember that second-cut hay has a higher nutritional value than first-cut.

As with everything else goat- and farming-related, it takes time, observation and a willingness to learn. Until you've developed these skills, you'll have to rely on the word of an experienced farmer, veterinarian, feed rep or neighbor, or you'll need to pay for a feed analysis. Remember, too, that if you don't know how to read a lab report, you'll also have to hire a nutritionist to interpret the information for you.

Nutrients for Life

All animals require essential nutrients for life. Here is a list of these nutrients, along with a brief explanation of each.

Water

Without fresh water, a goat's body will not function, and she'll die in a matter of days. Goats will not drink from a dirty water source or from water that has an "off" odor, and they are not much interested in cold or icy water. If possible, offer warm water to your goats in the winter.

Many factors influence the amount of water a goat needs to drink, among them temperature, humidity, water availability in feed, feed intake, milk production, and the goat's breed, size and general health. Since the main component of milk is water, there is a direct relationship between milk production and the amount of water a goat drinks. A goat typically drinks three to five times a day, and since the act of milking stimulates the goat to drink, she almost always drinks just after milking.

Proteins

Essential amino acids are building blocks for muscles (meat), tissue, hair, blood cells, hormones and enzymes; they also play a part in milk production. Amino acids come from two sources: dietary proteins in feeds that are not changed or degraded in the rumen (these are called bypass proteins); and microbial protein, which is a by-product of the microbes in the rumen. Since all the necessary amino acids are created in the rumen, the composition of those acids in a protein found in feed is, for the most part, unimportant.

Fats

Fats are a source of energy as well as a source of important nutrients. Fats carry the fat-soluble vitamins A, D, E and K, and many essential fatty acids are critical for body function.

Carbohydrates

A goat's main source of energy comes from carbohydrates. The amount of energy available to the goat in a feed, whether grain or forage (hay, straw, and so on), differs according to the type and quality of the feed. The more digestible a feed is, the higher its quality, expressed in the phrase "total digestible nutrients", or TDN. Digestibility in this context refers to fats, proteins and carbohydrates. The younger the grass at cutting, the higher nutritional value hay has. As grasses mature, digestible fiber decreases while indigestible fiber

A raised, well-protected goat feeder doubles as a wall to the goat pen.

increases. Protein levels also decrease as grasses mature.

Vitamins

A goat's body depends on vitamins to fuel growth, repair and metabolic processes. There are two groups of vitamins: fat-soluble and water-soluble. Fat-soluble vitamins can be stored in fat tissues in the goat's body, while water-soluble vitamins are used as needed. Excess water-soluble vitamins pass through the goat's body via her urine. When nutritionists balance feed rations, only the fat-soluble vitamins and the B complex vitamins are typically of concern.

Minerals

There are two categories of minerals: major and trace. Major minerals include calcium, phosphorus, magnesium, potassium, sulfur, sodium and chloride. Trace minerals are needed in much lower quantities and include cobalt, copper, iron, iodine, nickel, selenium, zinc and manganese.

Do not offer goats a product intended for sheep—their copper requirements are very different. If there is no product available for goats, one for cows will work fine. Buy a product that has a 1:1 calcium-to-phosphorus ratio.

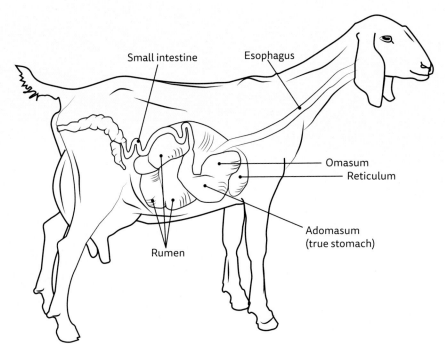

The intricate digestive system of the goat.

Fiber

Goats, cows, sheep, camels and deer are ruminants, a classification of animals that have a specialized, efficient digestive system, proficient at digesting fiber and turning it into fatty acids. In place of a simple stomach, ruminants have a complex organ made up of four different compartments designed to digest roughage (food that is high in fiber)—usually grass, hay or silage. (Silage is fermented fodder made for ruminants from grass crops or corn.) The four compartments are the rumen, the reticulum, the omasum and the abomasum, and each one plays a different role in the goat's digestive process.

Ruminants famously "chew their cud," though not everyone is clear on what that means. In goats, the process of rumination can start as early as four weeks of age and definitely is in place by the time a kid goat is weaned from milk. As a goat chews its food, it starts to break down the roughage, which becomes soaked with saliva and then swallowed. This softened mass of food is called the "cud." After it is swallowed, it enters the rumen, where it is broken down further by microorganisms that aid digestion. As the goat "ruminates," the cud is brought back up its throat to the mouth, where the goat chews it again before re-swallowing.

In other words, when you feed your goat, you're not simply feeding the goat—you're also feeding the rumen's complex ecosystem of bacteria. These bacteria consume the nutrients fed to your goat and produce by-products of essential nutrients. It is these bacteria-generated nutrients that the goat absorbs into the bloodstream. A healthy, well-functioning rumen is essential to a goat's health.

Once again, careful observation of your animals is the key to their health. As your goat chews, watch its neck. You should see the goat swallow and then chew again, at regular intervals. Occasionally, a goat will belch to release the gas that builds up in its rumen during what is basically a fermentation process. Make a note of when your goat chews its cud. Our goats tend to chew theirs when they are resting and relaxed, often when they're lying down inside the barn or resting out on pasture. They also chew their cud when being milked. For me, that's a good indicator of goat comfort. If a goat is chewing her cud while being milked, I know she is a happy and relaxed animal.

A Saanen doe enjoying her hay. Hay is an essential part of her daily diet.

If the rumen stops working as it should, whether as a result of indigestion caused by improper feeding, overeating or bloat (the goat's inability to release the methane gas that naturally builds up in the rumen), you must act quickly if you're to keep your goat alive. If you suspect a problem because your goat's eating behaviors change or if the animal is grinding its teeth, twitching or has a fever, call in your veterinarian.

Good Rumen Function

To keep the rumen bacteria communities thriving, follow these eight rules when feeding your goats:

1 Provide lots of clean, fresh water

One of the factors that drive milk production is water intake. At our farm, we always say that a generous supply of fresh, clean water is like getting milk for free. Since water is also essential for healthy organ function, goats of all ages need lots of it. Goats need extra water rations when the weather is very hot or dry, and when a female goat is pregnant or lactating. In the winter, warm water will encourage your goat to drink more; water that is free from ice is essential.

2 Remember the importance of long fibers

Long fibers in your goat's diet are critical to keeping the rumen healthy, since the bacteria in the rumen need a place to attach themselves. Your goats need an ample supply of forages in their diet, and grasses from pasture, hay and straw are excellent sources.

Straw, which is the stalk that is left over from grains such as wheat and oats after harvest, does not have much nutritional value, but is an excellent source of fiber. Fresh spring pasture is a great source of nutrition, but is very low in fiber. If you're going keep your goats on pasture, where free-choice feeding is an option—which means that your animals have access to a range of feeds and can balance their own diet—straw combined with pasture is an excellent way to keep those rumens functioning. Goats must have access to both grass and straw at all times. If your goats are like ours and are afraid of the dark, or if predators are a worry, it's fine for them to go without grass during the dark night hours. Just make sure they have straw available. You may also wish to supplement with hay in the barn, but keep in mind rule number 3.

3 Do not make drastic changes to your goat's diet

The types of bacteria that grow in your goat's rumen are dictated by the kind of feed your goat is eating. For example, you may be offering a dairy-grain ration to your goats. One day, you run out of that feed and can't get to your feed mill for several days. Since you have a bag of oats on hand, you decide that oats are better than nothing.

Meanwhile, your goat's rumen has a healthy supply of bacteria that are used to eating the dairy-grain ration. All of a sudden, that feed source disappears. Without the dairy ration to sustain it, the bacteria population that lives on this ration decreases drastically. Now, however, there are oats in the goat's diet. Floating around in the rumen are bacteria that like the oats, and the population of this type of bacteria then increases. Once the population is of sufficient numbers (usually after three to seven days), the digestibility of the oats increases, and your goat starts to receive the full benefit of the new feed. But the next week, when you find the time to make a trip to the mill for a fresh supply of your regular dairy ration, the bacteria population has to go through yet another change. This variability in diet significantly decreases the efficiency

Signs of Protein Imbalance

If your goat's stools are loose and coming out as soft piles rather than round pellets, there's a good chance her diet is too rich in protein. If your goat has a rough, coarse hair coat with no shine, there's a good chance she's lacking in protein. Changes in your goat's stool are seen quickly, while changes in her hair coat can take weeks, even months, before a difference is noticed. Remember, minimize feed changes and, when a feed change is made, small, subtle changes work best in order to avoid upsetting the delicate balance of microbes in the rumen.

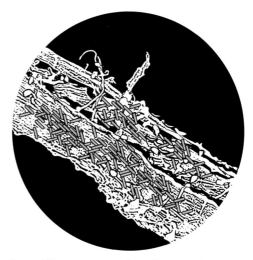

Long fibrous matter, from hay or other roughage, is necessary in a goat's diet to provide a habitat for rumen bacteria.

of your goat's rumen function. What's the moral of the story? Humans may not like to eat the same thing day after day, but your goat does.

4 If you feed grain to your goats only once or twice a day, make sure they can't gorge themselves

Feeding hay before grain is a must. This way, the goat's rumen is supplied with the long fibers necessary for proper digestion. Goats also eat less—and more slowly—when they don't have an empty rumen. Feeding goats during milking times is one way many people prevent their goats from gorging on grain. When your goat is tied up or locked in her head gate, she can eat only what's in front of her. However, multiple goats in a pen can be a problem at grain-feeding time because often there's one goat who's queen of the pen and bullies her way to the bulk of the grain feed. This bully not only puts herself in danger of bloat and other metabolic disorders but also leaves the other goats in the pen missing out on their ration of grain.

At our farm, we've found our goats butt each other and fight just as much when they're side-by-side as when they're face-to-face. Our solution is to provide more rather than less feeding space; a feeder in the middle of the pasture offers more available eating spaces. If there's lots of room at the trough, the weaker goats have an equal opportunity to access grain. A feeder constructed high off the ground with a bar designed as a footrest allows goats to fulfill their natural desire to elevate themselves while eating and helps keep feet out of troughs as well. The height of the footrest and feeder should be dictated by the size of the goats.

5 Remember that a goat requires a different level of nutrition at each stage of its milk production

For example, a young growing goat or a lactating goat needs more protein than a mature goat that is neither growing nor producing milk. For milking goats and growing kids, it's common to supplement with a 14 to 20 percent protein grain ration, depending on the quality of forages available (higher-quality forages = higher levels of protein). A goat that is on high-quality browse land with a variety of plants actively growing may not need any supplementation at all. Since goats are not a productive species when on pasture (in other words, when they're just eating grasses), a grain supplement will most likely be needed.

6 Always give your goats the best-quality feeds available

Do not feed them dusty hay, and be meticulous about cleaning old feed out of feeders before adding fresh feed. Contrary to popular belief and cartoons, goats do not eat tin cans and other garbage. Goats are actually the pickiest species of animal I have ever fed. While they do love to eat the bark and needles off our pine trees, as well as the geraniums in our flower boxes, they consume only good-quality feeds. Those trees and geraniums were the picture of health before the goats got to them.

7 Know your minerals

Copper, zinc, calcium, phosphorus, magnesium, potassium, sodium, iron, cobalt, iodine, selenium, sulfur and manganese are common minerals present in purchased, prepared goat mineral mixes. There may be vitamins added as well, such as A, D or E. Some mineral products also contain fat and protein and usually come mixed with molasses in a lick format or a thick liquid. Depending on your situation, these products may be a good alternative to supplementing with a grain ration. In the Great Lakes Basin, where our farm is located, the soils are very low in selenium, so it's important for our goats (and all animals) to have this mineral added to their diets. At the same time, selenium can be toxic in large quantities, and the difference between just enough and too much selenium is very small. As a result, feed companies in both the United States and Canada must follow strict regulations for adding selenium to feed and mineral mixes.

For animals that are post-weaning, always provide free-choice access to salt and other minerals. I strongly recommend purchasing a mineral product that's specially formulated and mixed for goats. The manufacturer has the knowledge, tools and resources necessary to prepare a product that provides the correct amount of minerals necessary for a goat to stay healthy. Talk to a knowledgeable staff member at your local feed mill or feed store to discover the best option for you and your goats.

As a general rule, if your goat is on a purchased grain ration, minerals should already be present and an additional supplement is unnecessary. In this case, salt can be offered in the form of a salt block. If your goat isn't on a purchased grain ration, minerals should be fed according to the manufacturer's recommendations with or without a salt block, depending on the available salt in the mineral mix.

8 Talk to your feed rep

If you want to use nutrition as a tool to push for better milk production, your best bet is to talk to the feed rep at your local feed mill. He or she will probably have more experience-based advice than your veterinarian, unless your vet has a specific background in goat nutrition. I also caution against doing much nutrition research on the Internet, unless you're certain the information is coming from an accredited ruminant nutritionist.

Feeding Strategy

Here is an example of my feeding strategy with a doe weighing between 36–45 kg (80–100 pounds) on a maintenance ration. She isn't producing milk, nor is she in the last trimester of pregnancy, so additional nutrition isn't necessary for milk production. She may be in the dry-off phase of milk production. A maintenance ration encourages the doe to stop producing milk.

- 2 kg (4 pounds) hay (approximately one flake per day)
- Free-choice water
- Mineral given according to manufacturer's instructions
- Salt block

This strategy can be used as a starting point. You can make adjustments, but keep the following in mind:

If the same doe is on grain, she will eat less hay. She may eat more or less hay, depending on the quality of the hay. She will eat more hay in colder temperatures (–10°C/14°F in a dry/draft-free environment).

If the doe's body condition score (BCS) increases, a cutback in hay is necessary.

If there is a lot of wastage, either the goat is being fed too much hay, or the quality is poor and she's leaving what she doesn't want to eat. Assess the hay left behind. If it smells or looks

Table 1: Body Condition Score

Score		Description	
1		Spine sharp, back muscle shallow	Lean
2		Spine sharp, back muscle full, no fat	
3		Spine can be felt, back muscle full, some fat cover	Good condition
4		Spine barley felt, muscle very full, thick fat cover	
5		Spine impossible to feel, very thick fat cover, fat deposits over tail and rump	Fat

Feeding Notes

- The term "forages" covers a wide variety of feedstuffs that supply nutrients and the fiber essential for rumen health. Forages include pasture, dry grass or hay, fermented grass (called haylage or baleage), fermented corn stalks and grain corn (called corn silage), and dried grain stalks or straw. For most readers, the term will mean dry hay or pasture and straw—feedstuffs that are user-friendly for small herds of goats because you don't require big equipment to move them.

- Fermented feeds are produced when the plant is harvested at a moisture content that is much higher than dry hay. This "wet" harvested material is wrapped in plastic or packed tightly in a silo to provide an anaerobic environment. The bacteria and microbes that grow on the feed break down and ferment the hay or corn, creating a product that can last a long time if stored properly. Fermented feeds are not eaten quickly enough by small herds, however, so they're not usually an option for hobby goat farmers.

- The most important quality to look for in your forage is an absence of dust and mold. This is not a concern in the pasture, but with dried forages, mold can be a big concern. Dust in hay is generally an indicator of mold. If a fine powdery white or gray dust plumes out when a bale is pulled apart, that isn't a good sign.

- Consider the maturity and quality of the forage. Hay should have a green color. If the hay has been rained on or left for days in the hot, drying sun, the nutrients will be diminished, and it will have a brownish look to it. Stalks of grass should be pliable and not too mature. There should be noticeable leaves left on the stalks.

musty or moldy or is very stalky and tough, the goat is simply being selective and eating only what she considers acceptable. If the discarded hay looks just as good as the original, smells sweet and has no off odors, she's probably just getting too much hay and a cutback is necessary.

If you see signs of discomfort such as teeth grinding, kicking at her abdomen or standing and "doing nothing," and there is no hay available, she needs an increase in hay.

Use your best judgment. Get to know your goat. By watching her body language and actions at feeding time, you'll know if she needs more hay or not. If your goats appear to be ravenous at feeding time (rushing to the feeder, bleating loudly, frantic to get that first bite), she'll most likely need an increase. A bale of hay naturally breaks into different sections, and these sections are called flakes. Our general rule for feeding goats is "one flake of hay per goat per day." One flake of hay from a standard 23 kg (50-pound) bale that contains 12 flakes works out to approximately 2 kg (4 pounds) of hay per day. However, remember that not every bale breaks the same way and not every flake is of standard weight.

Remember, feeding goats well is truly an art based on knowledge. The more goats you feed, the better you'll become at feeding. The key is to be a good observer. Slow down and take time to watch your goats, their feed and their environment.

Life Stages Diet

In the first 24 hours of life

A newborn kid requires only colostrum, the first milk produced by the maternal doe. Colostrum is rich in fats, proteins, milk sugar and the antibodies necessary to jump start a newborn kid's immune system. It also acts as a laxative that stimulates the bowels of a newborn kid.

A kid in the first four to eight weeks of life

Use only milk and hay on which to nibble. Although grain is not a necessity while kids are on milk, it is a benefit to growth. Introduce grain a minimum of one week before weaning the kid from milk. It's preferable if the grain offered is a starter ration—a very digestible blend of grains, minerals and vitamins that has been combined with essential fatty acids, yeasts, molasses and, sometimes, milk ingredients to encourage intake by the kid and stimulate early rumen development. We prefer a pellet (rather than a textured ration) that provides complete nutrition. A textured ration allows kids to pick and choose what they eat out of the mixture. Typically, they eat the corn and other grains and leave the vitamins and minerals behind, much like children leaving their vegetables on a plate. A pellet forces the kid to eat all the components in the ration.

Kids during weaning

Use grain, preferably a starter ration, and hay. If a kid is still nursing, consider creating what is known as a creep feeder space (a quiet space in the barn that is warm, well lit and away from other animals), where you can feed the kid grain to encourage weight gain and weaning.

Doe and buck kids over four months

Use free-choice hay or browsing and a free-choice mineral. Grain is not necessary but could be an asset, depending on hay quality and desired rate of growth. The higher the level of nutrition, the faster the kid will grow. If fed properly during this stage, females will give more milk as adults. The key is balance. Feed enough high quality feed, including grain, so the doe kid grows to maximum potential but does not get fat. Fat doe kids end up being poor milkers as do under-nourished doe kids.

Does at breeding age

No additional nutrition needed; place on a maintenance ration.

Breeding bucks

Very high-quality forages or a grain supplement is necessary. If the buck is breeding only a few goats, a maintenance diet (see page 38) is sufficient.

Does in their last three to four weeks of pregnancy

Need grain supplementation to support the increased nutritional demands from the unborn kid and to ready them for birth and milk production. However, before this, pregnant goats need only a maintenance ration.

Fresh goats (goats that have just kidded)

Use the best-quality forages you have available and a grain ration that is specially formulated for dairy goats. This diet will allow the goat to produce milk to her full potential. Feed is only one part of the milk production equation—goats also need sufficient clean water—but if the nutritional building blocks are not present, the goat cannot produce the milk.

Lactating goats

Use high-quality forages and a minimum of a high-quality mineral. I highly recommend feeding a dairy goat grain ration as a supplement, especially if you want a higher milk yield from your goat.

Goats at the end of their lactation

Decrease or withold the grain ration. The most common problem with goats during this drying-off period is weight gain. As milk production decreases, the energy that previously went into milk production is deposited as fat.

CHAPTER 3

Breeding Your Goats

ONCE YOU'VE PURCHASED a female kid, the next challenge is raising her to breeding size. To produce milk, a doe must be bred. The average doe reaches breeding size when she's about seven months old and weighs 32 to 36 kg (70 to 80 pounds) or roughly 70 percent of her adult size.

Breeding is probably one of the biggest challenges in keeping goats. Most female goats are seasonal breeders. That means they typically ovulate, or come into heat, at a particular time of year, usually as the days get shorter in late summer and fall. In our 75-goat herd, 98 percent of our does come into heat during the fall; this drops to only 30 percent that come into heat naturally during the winter and to less than 20 percent during the spring.

Of course, it's possible to breed does at other times of the year, but the likelihood of success isn't nearly as high. If a doe is bred in the very hot summer months as opposed to the autumn, for instance, there is a significant probability that she will absorb the embryo during the early stages of her gestation. Similarly, while does may continue to cycle from January through March, you're much less likely to see a heat during that time of year.

The breed of goat and the weather also play a big role in ovulation and a doe's ability to carry a pregnancy, as does the doe's physical condition. On our farm, does are ideally in perfect body condition at breeding. Fit, healthy does have the best chance of maintaining a pregnancy. A doe in prime condition, with a shiny coat and bright eyes, is likely to produce more ova, increasing the likelihood that she will deliver two kids rather than one. And that, in turn, means she will produce more milk.

Following a Schedule

The brief period when a doe will "stand" and allow a buck to breed her (also called a standing heat) is roughly 12 to 36 hours. Once bred, the doe has a gestation period of some 150 days, or five months, and she usually gives birth when the days grow longer in spring and early summer. In our experience, September 1 through December 31 is prime breeding time for goats. January is hit-and-miss. As a result, our kidding season takes place from February 1 through May 31, and, for a higher chance of successful conception, we plan our breeding, kidding and milking schedules around these dates.

By following these natural markers, we've created a 365-day cycle which includes two months of rest that leads to a healthy rate of reproduction and maximum milk production. This is also the preferred schedule followed on most commercial dairy-goat farms. Expecting a high-milk-producing doe to reproduce more often than once a year puts her under a lot of stress.

If you extend the reproductive cycle for more than a year, however, you then run the risk of slowing the genetic potential of a herd. When fewer offspring are born, fewer young animals are then raised to replace the older goats in the herd. Assuming that we are raising healthy goats, we always expect that the next generation will produce even more impressive kids than those in the preceding one. For these reasons, it's important to keep replenishing and strengthening the herd.

In the days before the doe delivers, her mammary glands begin to produce a substance called colostrum, a thick, yellowish low-fat milk containing immune cells and antibodies that protect the newborn from disease in the first days and weeks of life. It is critical that the kid is fed this nutrient-rich form of milk during its first 12 to 24 hours. It takes a few days for colostrum to transition to the goat milk that we humans wish to consume. This transition milk is rich in fats and proteins and is best fed to your goat kids. After the sixth milking, or two to three days after kidding, milk can be taken and used for human consumption.

Breeding Your Goats

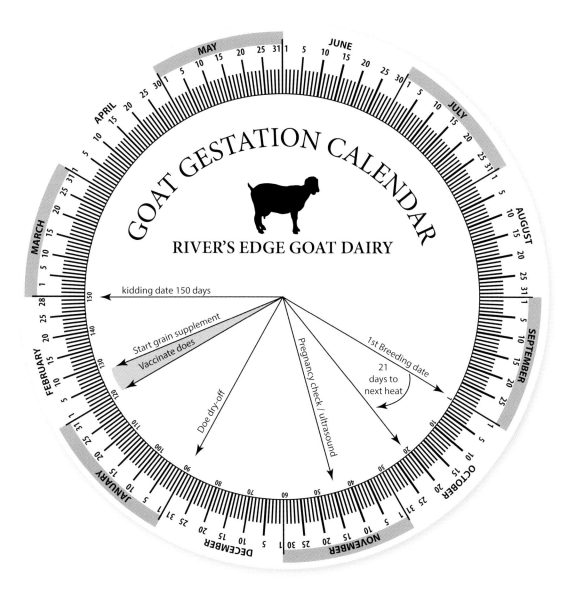

A breeding wheel consists of two wheels, one spins on top of the other. The breeding date arrow is turned to the date that breeding was observed or expected. The remaining corresponding arrows give dates of necessary actions to keep goats at optimum health during gestation. Conversely, the kidding date can be picked and the date necessary to breed will be known.

A healthy doe goes on to produce regular milk at a good rate for roughly another 305 days. Farmers don't want to interrupt milk production that covers the costs of labor and feed, nor do they want to continue milking a dairy goat that is producing small amounts of milk, which is what naturally happens after the 305-day mark, depending on the breed of your goat, the time of year and how well the goat is managed.

At the 305-day mark on our farm, we give our does a 60-day respite from milk production. It's best to let a goat go through her natural cycle of a decline in milk production and a rest before she kids again. The process of stopping milk production is called "drying off," which we discuss at the end of this chapter.

You're not obliged to follow this particular schedule, but it is important that you plan and follow some schedule. The length of time your goat produces milk will vary according to her breed, the time of year and how well she's managed. The important point to note is the time of breeding. In a 365-day cycle, where a doe kids once a year, the goat must be bred after seven months of milking. If a goat kids March 1, she should be bred October 1 so she kids March 1 the next year. If you don't have your goat bred sometime around October 1, you run the risk of not having milk the following spring.

Some does produce milk for more than a year while others produce milk for only eight months. But until you get to know your goat and her milk production, I recommend you stick to a 305-day lactation with a 60-day rest. It is critical that your goat has this two-month period of rest to rebuild her reserves and get prepared to give birth.

Breeding Basics

Simply put, your doe has to be with a buck in order to be bred. A number of strongly held views exist on whether a goat farm needs to have a buck as a permanent resident. Feeding and caring for a buck year-round and managing his access to ovulating does are big responsibilities and not to be taken lightly. If your herd is small, having a buck installed on the premises may not make financial sense.

Furthermore, a full-grown buck can be quite hard to handle, if only because of his size, and often requires stronger penning than a doe would. He may also be physically aggressive because he has not been raised with the same level of personal handling as we give our milking goats.

The biggest drawback to keeping a breeding buck, however, is his odor. A breeding buck secretes oils. Female goats love that smell. Humans definitely do not. (A buck may also urinate on himself when excited by the presence of a doe in heat, which contributes to his nasty scent.) These oils get on everything the buck touches. If you so much as touch a gate or fence the buck has been leaning on, you're a carrier. And if you're not careful, you will continue to spread the oil on everything you touch, from your clothes to your furniture. This oil is also very hard to wash off—goat handlers have tried everything from toothpaste and pet shampoo to hand soap and coffee grounds. It sometimes takes a good 24 hours to get rid of buck taint. And the smell only gets stronger during breeding season. Some goat handlers recommend giving long-haired bucks a midsummer clip, and providing clean bedding and shelter from the elements also helps. Some of us set aside a special set of clothes for buck handling.

Because of the costs of feed and care, many goat owners decide that it isn't practical or economical to keep a buck around all year just for the purpose of servicing one or two goats.

If, however, you're intent on purchasing a buck, you have a few options. Consider buying a young buck of seven months to one year of age. This buck is old enough to breed, although it may take him more than several tries to figure out what to do with a doe in heat. Once your doe has been bred and you're confident she is pregnant, you can either sell or butcher your buck for meat, since a buck at around a year of age is young enough to yield excellent-quality meat. You could also sell your buck to another farm for breeding.

Alternatively, you could buy an older buck, one that is an experienced breeder and may have some milking daughters available for viewing as evidence of his genetic profile. A mature breeding buck costs more than a young buck, and a proven breeding buck (one whose daughters show excellent conformation and production) costs even more.

Finally, if you live in an area that has a number of goat farms, consider contacting other interested parties and setting up a buck-sharing program whose participants trade bucks from farm to farm. It's a system that works well, provided there is trust among all participants. Each herd must be in good health, and there should be a shared interest in strengthening the herd's genetics every year.

Breeding Options

If you decide not to keep a buck, you have three breeding options.

1 Artificial insemination (AI)
Although AI is becoming more common for goats, it is still a very expensive breeding strategy and the conception rate is quite low when compared with natural-breeding results. However, AI does offer two significant benefits. First, since semen is collected only from breeding bucks with good genes, these genetic strengths are passed along to the next generation of goats on your farm. Second, AI eliminates the potential for disease to be passed from buck to doe.

"Buck-in-a-jar"

Some people use "buck-in-a-jar" or "a buck rag" to bring a doe into heat. Find a neighbor who has a buck that is ready to breed. Rub a rag all over the buck to collect as much of his stinky oil as possible. Place this rag in a jar, bring it home and let the doe smell the rag. The scent may bring her into heat. Although I have not had this method work for my goats, goat-raising friends swear by it. The smell of a foreign buck may also stimulate a buck that is not yet ready to breed. The thought that competition is near is often enough for a buck to get his mojo going.

If there is a vet or technician in your area who performs this service, it may be worth the investment.

2 Send your doe to a farm that has a breeding buck on site
Although it is very easy to put your doe in your truck or car and drive her to a nearby farm to have her bred, there are drawbacks. One is that, if your doe is in milk, you will not be able to milk her unless you're able to take the time to visit the farm twice a day. If your goat is a doeling (under 12 months old) and has not been bred, that isn't a problem. Another potential problem is that your doe could pick up a disease in a different setting. Needless to say, farmers will be concerned about the possibility that your doe could also transfer disease to their animals from yours.

Another problem may be the lack of information about the breeding buck's production history. Not many farms keep production records on their goats, and even fewer calculate production potential for a buck. Unless the buck in question has a record of producing several milk-producing generations of offspring, you can't know for certain the quality of his input.

A doe in standing heat being bred by a young buck.

3 Bring a buck to your farm

If you're planning to breed goats naturally, this is probably your best bet. Once again, your primary concern should be about disease transfer. Borrowing a buck from someone may not be an option because of the risk of disease transfer. You don't want any disease from the buck, and the owners of the buck don't want their buck picking up disease from your doe.

In my experience, purchasing and re-selling or butchering a buck is the best option. Aim to purchase your buck at least one month before the date on which you hope to breed your doe or doeling. Keep the buck for a minimum of one month (or the equivalent of one heat cycle) after your goat is bred, just to be confident that the breeding has been successful.

If you have the room, keep the buck and doe in completely separate buildings. If that isn't an option, maintain the buck in a separate pen where even nose touching isn't possible. If you don't have any extra space, you'll have no choice but to put your new buck in with your doe.

Keep in mind that the buck may be carrying diseases even if he shows no signs or symptoms; the same is true of your doe. If you're able to separate the animals, observing the buck for three to four weeks before breeding takes place should be sufficient. The goal of this separation period is to watch for signs of a potential disease which are visible to the human eye. Look for skin lesions such as abscesses or warts, coughing or sneezing, weight loss or just poor health in general. If any of these signs appears, call your vet for best advice.

Part of the mating ritual: Does flag their tails to send their scent pheromones to the breeding buck with the intent of enticing the buck to breed.

Time to Breed

If your doe hasn't come into heat naturally, you can use a few methods to encourage her. Sometimes introducing a buck in full breeding (read: very smelly) to your female goats does the trick. This process usually takes about two weeks but, for it to work, it is important that the doe doesn't see or smell the buck for several weeks or months before the introduction. It's also important that the buck is ready for breeding.

Bucks are also seasonal breeders and respond to the shortening of the days as does do. We once sold goat milk on a commercial scale and needed year-round production. We brought 30 goats into heat artificially by using hormones. When we introduced the buck to the does, he walked up to the manger, ate and then proceeded to have a nap. Clearly, this buck was not in the mood to breed. Needless to say, we had a bunch of frustrated female goats. We weren't too happy with the situation either.

What now?

One to two weeks before the date you want your doe bred, introduce your buck to your doe. Simply put them in the same pen. They will figure out the rest. Observe your goats carefully.

Several behaviors signal your doe is in heat. Her vulva becomes red and swollen, and there is a clear mucous discharge. She is more active, and her urination is more frequent. She is also likely to vocalize a little more, and her appetite and milk production may both decrease.

One of the classic signs of a doe in heat is wagging, or flagging, of her tail. If a buck is present, she will also rub herself and make herself available to him; conversely, the buck will be very interested in her. Depending on your goat's personality, she may also become more affectionate with you.

When a goat is in standing heat, she is ready to be bred. This is the point at which she will stand for the buck for breeding. She will also stand quite firm and square if you put pressure on her back with your hands. When you observe these signs and the buck is in with your doe, observe their actions carefully. Ideally, you'll witness the breeding as it takes place and be able to record the date in your journal. Then make a note to observe your goat in 18 to 22 days, the time between ovulations during breeding season. If your goat is indeed pregnant, she will not show a heat in 18 to 22 days from your observed breeding. If she does show a heat, the buck should breed her again. Follow the same note-taking procedure. If you don't see a heat, you know your doe is bred and her kidding date will be approximately five months from this date. Record that date as well.

When you're confident your doe is bred, it's time to make the decision about whether you're going to keep your buck, sell him or take him to the butcher.

If the doe you have just bred is a doeling and she's not in milk, just sit back and enjoy the time with your goat until she kids. Practice handling her, putting her in a milk stand or another suitable place for milking, and rub her teats (where her udder will be). Getting your goat accustomed to this handling slowly over a long period will make your job—and hers—much easier and less stressful when she reproduces and starts producing milk, a phase is known as "freshening."

If Your Doe Is Mature

If your just-bred doe is mature and already in milk, don't let breeding or the fact that she's bred keep you from milking her twice a day. You may see a slight drop in milk production when your doe is in heat. That's because the energy normally used for milk production is being used elsewhere. Your goat may not rest as much or eat as much when she's in heat. Once she's bred, she will return to old habits of eating and resting, and her milk production should return to former levels. Keep milking your doe until 60 days before she's due to kid or until she naturally stops producing milk.

How to Stop Your Goat from Milking

Your goat may dry off naturally around 60 days before kidding, or even earlier. Her milk production may gradually decrease over time after she's bred. If this happens, great—go with it and remember to decrease feed accordingly. But if your goat is still producing a large volume of milk 90 days before kidding, you should start thinking about milking your goat once a day rather than twice. That reduction should encourage her to slow down her milk production. As your goat's production drops, you should also decrease her feed. If your goat is still milking heavily, decrease any grain ration or remove it completely to slow her production.

If your goat is still milking heavily (½ to 1 liter/quart, or more, per milking), you may wish to consult your vet. If you stop milking a doe when she's still producing a decent quantity of milk, the incidence of a goat freshening (coming back into milk production) with mastitis is increased. Be warned about the severity of a goat freshening with mastitis. It can develop into gangrene, which can be fatal.

CHAPTER 4
Kidding

THE MOST REWARDING part of keeping goats is experiencing the birth of a new kid. Thankfully, the process of kidding is usually straightforward. But every once in a while, it doesn't go quite so smoothly and an intervention is necessary. That's when calm composure is an asset. First, let's review what happens when kidding proceeds normally.

Although does should always be inspected twice a day at feeding time, as the 150-day post-breeding date, or kidding date, approaches, observe your bred doe more frequently. Like humans, goats rarely give birth on the exact birthing date, so start watching very closely about a week before the big day. If your doe has not kidded before, she's probably quite active and exhibiting a normal appetite. Typically, she'll have just one kid, but if your doe has already kidded, she's most likely pregnant with twins, triplets or more. She may lie down and rest more, she may groan and her appetite may suffer, if only because she doesn't have space for feed.

Kidding kit
- Clean towels and rags
- Close-fitting nitrile gloves
- Shoulder-length plastic gloves
- Lubricant
- Iodine for navel dipping
- Heat lamp
- Sharp scissors
- Goat collar
- Twine or butcher string
- Lead rope
- Plastic bottle and nipple for bottle-feeding kids

Labor

Labor unfolds in three stages:

Stage 1
Kids move into position, and the doe's cervix starts to dilate. Watch for several early signs. Often, the doe picks a nice quiet corner to rest. If there are other goats with her, she may become aggressive toward them, though they may also give her the space she needs. The area around the base of the doe's tail may become softer as ligaments and muscles relax. The doe may go off her feed though, in our experience, does often nervously nibble at hay just before kidding.

Stage 2
Active labor. The doe is actively pushing, striving to deliver her kids. During this stage, it's best to watch quietly from afar and give your doe space. Most does like their privacy and won't start pushing until they feel they are alone, though it's also true that some don't mind having company. In our experience, once a doe starts into active labor, she does not stop until all her kids are delivered. So once your doe has progressed well into the active phase of kidding, by all means stay and observe.

Stage 3
Delivering the placenta. Each kid has its own placenta, a shiny, reddish, jelly-like mass with white cords running through it. So it is important to keep an eye on the doe until all kids have been born and all placentas have been delivered. Most does deliver the placenta quickly, usually 20 minutes to four hours post-kidding. However, if even a remnant of the placenta remains in the goat's uterus, infection and even death can follow.

Animals do not feel pain in labor as humans do. After watching hundreds of animals kid, the only ones I have seen in pain are those that have a particularly difficult birth. All animals release pain-suppressing hormones during the birthing process. No pain medication is necessary, nor do I recommend any, even for a difficult birth.

Humans view pain as something negative. Animals do not. Pain is what keeps our animals from running on legs that cannot support running. Pain keeps a doe that has just kidded quiet and resting. By taking pain meds, a doe may exert herself more than she should, potentially damaging her body. After giving pain meds to a goat, we cannot tell her to rest and take it easy for the next few days. Pain meds will make the goat feel better and her actions will reflect how she feels. This response isn't good for the goat, but the human feels better.

A goat that is in too much pain will not get up to eat or to look after her kids. If your goat is in this much pain, more damage has been done during the birthing process than the average goat owner can deal with effectively. If you feel that pain medications should be given, this is time to call your veterinarian. Most likely, the doe will need stitches and/or antibiotics.

If your doe received proper nutrition, she needs no special treatment after kidding. Your doe was receiving good-quality hay and some type of grain supplement three weeks before kidding. Ideally, your doe is neither too fat nor too thin; it is better to have a doe on the thin side rather than overweight. An exceptionally fat doe is susceptible to pregnancy toxemia: the doe burns fat too quickly with the exertion of kidding and essentially poisons herself. This condition is difficult to treat in goats, and most often the result is death. As with all aspects of raising goats, management, maintenance and preventive steps are crucial.

Kidding

Make a clean, dry bed of straw on which your doe can kid. Do not use wood shavings, which are too fine and dusty and may interfere with the newborn kid's ability to breathe.

Kidding typically progresses uneventfully with no need for outside help. Sometimes, however, the doe needs assistance, and if you don't provide it, there's a good chance both the doe and the kid could die.

The tricky part for the goat handler is recognizing when the doe needs help and when it's best to sit back and wait. As always, the extent to which you know and understand your goat is what makes the difference.

A goat's gestation is roughly 150 days, or five months. You should have an idea of when your goat was bred and, therefore, a sense of a normal delivery date. It's impossible to predict exactly what a doe will do before she kids, but a few behaviors are commonly seen. The first sign is the goat "bagging up"—the udder starts to fill with milk. On a doeling (a first-time kidder), the udder is most likely quite small, but you can observe something more than just teats. (We have seen goats bag up three weeks before kidding and just after kidding, so this is by no means an ironclad rule.)

A doe about to kid often creates her own space, and the rest of the herd may avoid her. Some does make a bed or a nest. They may get up, turn around and lie down again and look as though they can't quite get comfortable. Some stop eating, while others nervously munch on hay. One thing that is relatively predictable is that a doe generally displays the same behaviors each time she kids. Keep notes. The information will be a great help to you the next time around.

Goats do not display "belly drop" as humans do. Gravity works in different ways on humans than on goats. The belly drop we see in humans is the baby moving closer to the birth canal. Since humans stand on two legs, not four, the woman's belly does indeed drop. In goats, the kid also moves toward the birth canal, this movement can be seen by the softening of ligaments in the tail head.

Sometimes the only sign that a doe has started the process of kidding is the appearance of a clear, mucus-like discharge from the vagina. Sometimes this discharge is extremely visible as a long strand; other times, it is hardly noticeable. If you miss it, the next sure sign that your doe is in labor is when she actively starts to push.

It should take no longer than 20 minutes for any one kid to be

Sac of amniotic fluid that surrounds kid is visible.

Sac is now broken and kid's hooves are visible.

First kid is born.

delivered. As the doe starts to push, the first thing you will most likely see is a round bag of fluids that looks somewhat like a red, veined balloon. This is the amniotic membrane that surrounds the kid. Sometimes this sac has already burst; in that case, the first thing you'll see is a pair of hooves. Sit back and relax. Leave your doe alone and watch for further developments, which can seem like an eternity.

Amniotic sac of second kid is visible.

Your doe may lie down or stand up during the active phase of labor. She may do a bit of both. Let her move freely as she wishes. Do not restrain her. She will not hurt the kid.

The doe continues to push until she delivers the kid's head and shoulders. Once this part of the kid has passed, the rest of the kid slips out easily. At this point, you can comfortably approach your doe and new kid. Remove more obvious bits of mucus from the kid's nose and mouth with a clean cloth so the kid can breathe.

Two hooves are clearly present.

The kid will sneeze out any residual mucus. If your doe has stayed put, move the kid in front of her where she can lick the kid clean. This action removes any lingering mucus and dries off the kid, but the licking also stimulates blood flow. The doe may appear to be a little rough with the kid. Don't be alarmed. However, if she starts stomping on the kid, head-butting it or pawing the ground,

Kid's nose and most of head is visible, along with two hooves.

move the kid to a safe, warm location. This is probably a sign that the doe is getting ready to give birth again and is looking for her space. If the doe jumps up and hurries to the other side of the pen, don't worry. Just make sure the kid is able to breathe, and leave as quickly as you entered. The doe will most likely return.

Hopefully, the second kid (and the third, fourth and fifth, if present) will arrive as the first kid did. Make sure each kid has the mucus wiped from its face as quickly as possible after birth. In between the birth of the kids, the doe may attend to one kid, then move off when she feels the urge to push and deliver another. Since goats don't mind if you touch their offspring, this is a great opportunity to dry off the first-born kids. Keep kids close to the doe but out of harm's way. Sometimes, although very rarely, a doe lies down on a kid and crushes it as she starts to deliver her next kid. After all kids have been born, the doe usually stands up and tends to them. At this stage, the kids are attempting to stand and are searching for an udder to get that first precious drink of colostrum.

Again, watch for the doe to pass her placenta. If it doesn't happen naturally, the action of milking the doe, whether by hand or by the kid nursing, will stimulate the uterus to contract and expel the placenta. This may take up

Doe continues to push hard as the rest of the head and shoulders are delivered.

Shoulders are out!

Once the shoulders are delivered, the rest of the kid slides out easily.

to an hour, but it often occurs quickly after birth. Sometimes, however, after a difficult kidding, a doe may not have the energy to expel the placenta. Watch and observe the doe until she has "cleaned" or passed all placenta because severe infection and possibly death can result from a retained placenta. Some goats eat the placenta, so unless you keep close watch, you won't know for sure whether the placenta has been delivered.

There may be a strand of dark pink tissue hanging from the goat's vagina, a remnant of the placenta. Do not pull on it because it may tear and leave a piece in the uterus. Placenta that is not passed by the doe, whether whole or a piece of it, is called "retained placenta." If your doe does not deliver the placenta on her own after four hours or so, it's time to call the vet.

Your doe will have bloody, mucus-like discharge for the next few days to weeks. The discharge starts off a red color, turns pinkish after some time and finishes off as a brownish color. If your doe produces any smelly or brownish, dirty-looking discharge shortly after kidding (12 to 24 hours), call your vet. There may be some placenta left inside the doe's uterus.

The essential act of a doe licking off her kid. If the doe does not do this, human intervention is necessary.

Both kids safely arrived. The visible sacs hanging from the doe act as weights to assist in the delivery of the placenta (afterbirth).

If the doe is relaxed, and doesn't mind human presence, the kids can be moved closer to the doe's head so she can lick them while having a short rest after delivering two kids.

If Intervention Is Necessary

If kidding does not proceed as described above, you may have to intervene. Don't panic. It's best to have someone with you if you think you need to assist the doe. I often calmly walk back to the house and ask my husband, son or daughter to help out. They change into warm barn clothes and walk out to the barn, and there's still time to assist the doe and save the kids. Often, the doe has sorted herself out by then, and the kid is lying by her side being licked clean. If you have a friendly relationship with an experienced goat or sheep farmer, keep that person's number on hand as well. Even if you're not comfortable about handling a difficult birth, I think this section is worth reading so you'll know what can happen. Remember, calling for help when needed is always recommended.

Sometimes a doe is in active labor with visual pushing, but no progress is being made. After 10 minutes or so, something should be visible. If not, put on a well-lubricated long glove and investigate. Gently enter the goat's vagina with your fingertips. Go slowly and carefully so as not to tear the goat's vaginal lining. Now you are looking for the goat's cervix, which should be dilated enough to allow you to insert your fingers and most of your hand. The doe's tendons and muscles will relax as you put pressure on the cervix. Keep gentle pressure, and slowly push your hand into the doe.

If the cervix is not dilated or you do not feel comfortable with this procedure, call your vet. You may choose to give the doe another 20 minutes or so to see if she can resolve the issue on her own. But if your doe has been trying to kid for longer than 30 minutes or you are not sure when she went into labor, place a call to your vet.

Here's how the kid should be lying for a normal birth, forelegs and head facing the vagina.

I'll now outline possible kidding difficulties, starting with the easiest scenario in which the kid is in a normal position but needs assistance. Remember, latex or nitrile gloves should be worn at all times when touching birth fluids, and that includes touching wet kids. Many diseases can be present in birthing fluids and the placenta. As well, you don't want to infect the doe or kid if your hands aren't clean.

If the sac, or amniotic membrane, is present, but the doe has not made any progression after 10 minutes, you must break the sac. It's not unlike popping a balloon. If you can't do it with your gloved fingertips, grab your sanitized scissors and give it a poke.

Sometimes this action gives the doe an extra bit of space to push the kid through. If you see part of the kid you didn't see before, great. Stand back and watch.

If no sac is present or if breaking the sac does not lead to progress, gently slide your gloved finger, with a bit of lubrication, inside the goat's vagina to feel for hooves. They will feel like sharp little bony projections. They are quite small, so go slowly and carefully, feeling all around as you go. Once you find the feet, give them a gentle but firm, steady pull until they are visible outside the vagina.

Once hooves are present, the next step is to find the nose. Very gently, slide your fingers along the legs and feel for the nose which should be sitting on top of the legs a short distance inside the vagina. If you have just pulled out the legs, use one hand to hold the feet, while the other searches for the nose. This will prevent the kid from pulling its legs back inside.

It is very important to be sure you have a nose in the birth canal befoe pulling. If you can't find a nose, do not pull. Stop and evaluate the situation to see if the kid is in a breech or head-turned-back position (see page 65).

When you find the nose, grab the kid's fetlock joints just above the hooves for a better grip. If your doe is standing, pull on an angle slightly toward the ground. If your doe is lying down, pull in the same direction relative to the goat, as if she were standing. Pull gently, but firmly. If it seems that the outside skin of your goat's vagina is too tight to let the kid pass, keep traction on the kid, but allow the skin to stretch for 30 seconds or so; then continue pulling slowly, but firmly. It is very important to pull at a downward angle. Sometimes, the legs need to clear the birth canal first, and then the head will follow. Pull the legs until the elbows pop clear. This should allow more space for the head. Your doe may help and push at the same time, which will speed up the process, or she may lie there and let you do all the work. She will most likely bawl. She may grunt. She's in pain. Don't stop—as soon as the kid is out, your doe will be greatly relieved. Remove the mucus from the kid's nose and reward your doe by placing the kid in front of her.

Breech birth

If you can feel feet but you can't feel a nose the kid may be coming out backwards or the kid is in a head-turned-back position. First determine if the front or back legs are present. If you're feeling its back legs, this is called a breech birth.

If you have a hind leg, find both feet and pull the kid out and downward.

The kid should slide out easily until you get to the shoulders. You should see a tail and hind end rather quickly. Once you're completely certain you have a hind end coming out first, pull relatively quickly. The kid will soon need its first breath of air. Once the umbilical cord breaks, your kid no longer has a source of oxygen from the doe. When the kid is delivered backwards, the cord breaks before the head is out—that's why it's important to deliver the kid quickly in this situation.

One or two legs back

If you do not see both front legs, you will need to assist the kid by finding and pulling both legs into the birth canal.

Put on a long glove, lube it well up past the wrist and try to find both feet. If one or both legs are back, you will have to insert your hand past the kid's head, through the birth canal, into the womb. Keep your fingers running along the kid's body. Take your time and try to identify each body part you feel: You're trying to run your hand down the shoulder to find a leg.

Once you find a leg, hook a finger around it and pull the leg through the birth canal, in much the same way you use a rug hook to pull a piece of yarn through a hole in canvas. Sometimes you can pull the kid out with just one leg present, but if it is a tight squeeze you'll have to go back in and find the second leg.

Once both legs and the head are present in the birth canal, you should be able to pull the kid out quite easily. As in the other scenarios, clear the mucus from the kid's nose and mouth, and place the kid in front of the doe.

Head turned back

If you have front legs and no head, you will have to find the head. This is one of the most awkward positions to correct.

For this procedure, ideally, you should have someone hold the doe. If there's nobody to help and she will not stay put, tie her up loosely by her collar. Put a long plastic glove over your latex or nitrile glove. Lube up your entire hand, past your wrist. Entering a doe's vagina must be done carefully to ensure you do not tear the lining.

Always start by following the kid's body. Place your fingertips on a leg and work in slowly and carefully. As long as you can feel the kid's body under your fingertips, you know you don't have any of the doe's tissue caught under your fingers. If the doe starts to push at any point, stop moving and wait until she relaxes. Try to "see" the kid with your fingers as you move your fingers along the kid's body.

Keep working around the kid's body

The many different birthing positions of kids.
A normal birthing position is always desired.

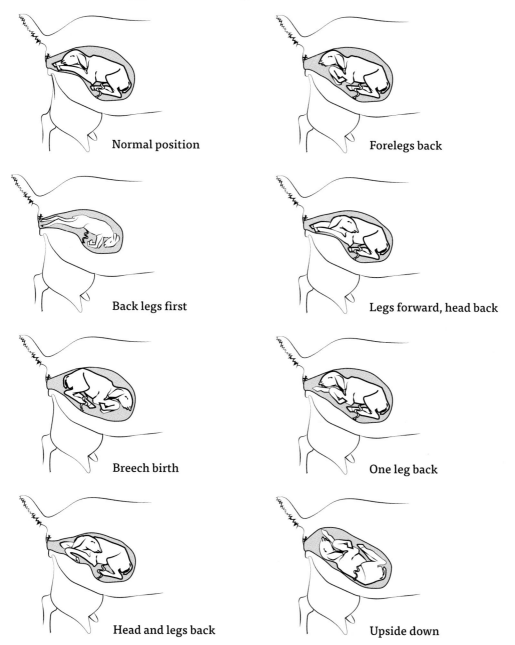

until you feel something familiar. Eye sockets, rib cage, ears and teeth are some of the more easily identifiable parts, your goal is to locate the head. During this process, the kid may try to slip its legs back inside the doe. If that happens, hold onto the legs with your other hand to keep the legs in the birth canal.

Once you've located the head, wrap your fingers behind the kid's skull. The palm of your hand should be resting on the top of the kid's head. Attempt to slide the head out into the birth canal. Your hand will be squished, and it will feel like you're squishing the kid's head. That's okay. If you can fit the head and your hand out through the birth canal, that's great. Continue pulling the kid until it is out.

If you can't move your hand and the head out together, get the head as much into the birth canal as you can. If you're able to get the head started in the right direction, it may continue on that path. Once the head is pointed in the right direction, you can also possibly hold firmly onto the kid's jaw and pull the head through the birth canal that way. Don't worry about hurting the kid—a kid is incredibly tough—and if you don't get the kid out, it will die.

If the head will not come through the birth canal, get the twine from your kidding kit. Fold the twine in half and make a loop at one end by running the loose ends through the looped end. You must put this loop around your fingers in such a way that you can insert the twine in the goat's vagina and around the kid's lower jaw. Once the twine is looped around the jaw, it should stay put as long as you keep firm and steady pressure on the twine with your outside hand. Wrap your inside hand around the back of the kid's head as you did before, and get the head started in the right direction. Once the head is pointing the right way, use the twine to pull the head as you slide your arm out. Your arm should be clear of the tightest part of the birth canal so that the kid's head will have room to move through it. Pull on the legs using the same pressure you're using with the twine. If no progress is made, the head most likely did not make it into the birth canal, or the twine has fallen off. Try again.

Multiple kids

In a normal birth, the kids should be lying side by side with front legs and head pointing toward the vigina, one kid should be further back in the womb.

If there are multiple kids, it's possible for the legs of two different kids to be in the birth canal at the same time. One brown foot and one white foot may be a clue that this is the case

When twins or multiples are present, the previous birthing positions apply, with the potential complication from a second kid.

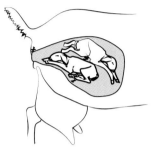

One kid facing forwards, one facing backwards.

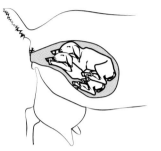

Two kids in canal.

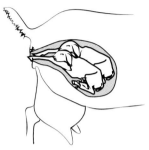

The head of one kid, with one leg, plus the leg of the other kid.

(though not a certainty, depending on the kid's coloring). Put a well-lubricated glove on and enter the doe and try to figure out exactly what body part is where. You will have to push one kid back and pull the other forward until only one kid is in the birth canal.

I have had some pretty interesting situations that have taken over an hour to sort out. In the end, the kids and doe were all fine, although exhausted.

As a doe is kidding, it is never certain how many kids she will have. Even an ultrasound is not always accurate. A delivered placenta, however, is a sure sign that no kids are left. If you haven't seen the placenta, one way to determine whether there are kids still to be delivered is to "bounce" the doe.

To bounce the doe, stand behind her, bend down and reach around her belly as though you're giving her a hug. Place your palms against the belly, directly in front of the udder. Pull up gently. If there's still a kid inside, waiting to be delivered, the belly will usually feel hard and uneven. (It may also feel squishy with a hard mass, which is the kid.) Move your hands up the abdomen, and repeat. Practice makes perfect when it comes to bouncing. It's all about becoming familiar with the feel of your goat's belly. The best way to know if your doe is finished kidding is to watch closely and carefully for the placenta(s) to be delivered.

Postpartum Kid and Doe Care

Note: This advice is relevant only for kids that are fed by their does. If your intention is to keep the kid off the doe for disease prevention, talk to your vet or ruminant nutritionist for instructions and information.

If your doe is in good shape, she needs no extra-special treatment post-kidding. Fresh water, good-quality hay and her grain ration will be all she requires. Warm water is more readily consumed than cold. Water should always be available to a goat in milk. Water must always be free of ice. The higher the water intake by your goat, the more feed she will eat and the more milk she will produce.

I don't recommend adding anything to your goat's water. Goats are ruminants. Adding sugar or an acid such as vinegar messes with the natural ecosystem of gut flora that lives in your goat's digestive tract. The absolute best thing your goat can eat post-kidding is her hay. The fiber in the hay keeps the gut flora happy.

Always be sure your goat has hay in her rumen before she eats grain. If you have the luxury of browse land for your goats, allow your doe free-choice access to it. The amount of grain you give to your goat will depend on the amount of milk she produces and her body condition score (see the section on body condition score in Chapter 2, *Feeding Your Goats*, page 38).

Once the kids are delivered, make sure they're warm and dry. If the doe is doing a good job cleaning each kid and the kid is sitting upright and looking bright, don't interfere. Sit back and watch to see what kind of care the doe provides. Some goats are fantastic mothers; others are not. Some goats feed four or five kids; others feed only one or two. Getting to know how your goat parents her kids will help you with your kid-managing decisions.

The Umbilical Cord

The umbilical cord almost always breaks on its own as the kid's hind end exits the birth canal. Alternatively the doe may cut it herself with a few bites. After kidding out hundreds of does, I have never had to cut a cord or have the doe do the job. The cord attaches from the kid's abdomen to the placenta, and the placenta stays inside the doe until after the kids are delivered. In a kid that's delivered backwards, or backside first instead

A doe standing protectively over her kids. Lots of dry, dust-free bedding is a must.

of head first, the cord also breaks after the hind end is delivered. This is why it's critical to quickly pull out a kid that is delivered backwards—a kid can suffocate when its head is still inside the doe and the cord has broken.

You'll know if the cord is too long if, when the kid is standing, the cord drags on the ground or the kid is stepping on it. A cord dragging on the ground increases the chance of picking up bacteria that will cause an infection. If the kid steps on the cord, there's the possibility of an umbilical hernia or excessive bleeding from the umbilical cord area on the abdomen itself. Never pull on the umbilical cord. The kid's navel should dry up within 12 hours—for sure, within 24 hours. If the navel has not dried up after 12 hours, I would highly recommend dipping it in iodine to speed up the drying process. If the kid's umbilical cord is still wet the day after birth, the pen is most likely wet and needs cleaning and dry bedding.

Meeting the New Kids

To meet your new kids, pick them up, give them a quick rub with a towel, do a thorough inspection to make sure all is well, and place them back down with the doe. This is a great time to apply a quick navel dip with the iodine to cleanse the remaining section of umbilical cord. Your doe and kid may complain, but there's no harm done.

Check to see if your kid is a buck or a doe. Check in two places: under the tail to look for a vulva, and between the kid's back legs for the presence of testicles. Both buck and doe kids have teats, so looking for them is not helpful. Sometimes the vulva is not easily visible, and sometimes the testicles are not fully descended. Laying the kid in your lap and flipping the back end of the kid upside down to expose the region between the legs is the best way to get a thorough view.

After delivery, kids should try to stand within five to 10 minutes of birth. (Does are almost always quicker to get up than bucks.) At this stage, the doe shows interest in her kids by calling to them, licking them, moving back and forth between all her kids, and making sure all are okay and accounted for. The doe is also getting to know the scent and sound of each of her kids—leave her alone to do her job. Kids may wander away from the

A group of kids under heat lamps on a cold −25°C day. Dry bedding and a draft-free location are essential if kids are to stay warm.

doe, call out to their doe and learn her returning call.

Within 20 minutes of birth, a kid noses around for an udder and the first precious drink of colostrum. The doe stands still, with her hind legs slightly apart. As the kid approaches her udder, a doe may nip at her kid. This is acceptable behavior. Once the kid latches onto a teat, these small nips encourage the kid to drink more vigorously. After all the kids have had a drink and all the little bellies are full, the doe lies down and the kids lie next to her for warmth. They all settle in for a much-needed rest. This is how everything should go in a perfect situation. But it doesn't always happen this way.

After (or maybe before) your kid

Tips—New Kids

- **Kids typically go to the shadows to find the doe's udder, so I recommend a well-lit kidding pen.** Not only does good lighting guide a kid to that all-important first drink of colostrum, it also helps you see what's going on and is appreciated by a vet who's on site to lend a hand.

- **If a kid is listless after birth, immediately take it to a warm, dry, draft-free location.** Rub the kid vigorously with a towel to get its circulation going. A kid's body-surface area is very big compared with the size of its body. Add moisture, and a kid can lose body temperature quickly and deplete its energy stores. Energy stores are lower after a tough birth, especially if the barn is cold and the doe had inadequate nutrition before kidding. I have often found that a cold kid will not get up or eat until it feels warm. But be sure that the kids are warm and not hot.

- **Healthy kids should be up and about, even bouncing around a few hours after birth.** When it comes to feeling good, temperature is key: Kids are most comfortable at 12°C (54°F) and up. If your barn is colder than that, consider putting up a sheltered area with a heat lamp or a heat mat.

- **Heat lamps and mats must be designed to be used in barns for agricultural purposes.** The colder the weather, the better the enclosure needs to be. On our farm, the coldest weather we experience is –25°C to –30°C (–13°F to –22°F). When the weather reaches these extreme temperatures, we need to put our kids in an enclosed area with a heat lamp at night. Healthy kids that received a full belly of milk can go for eight to twelve hours without any milk, provided they're warm.

has had its vital colostrum intake, you should see your kid peeing and pooping. The stance your kid takes to pee will confirm its sex. Bucks stand stretched out—and the urine may appear to run out of the navel. This is normal. Does will take a slight squatting position to pee, and the urine comes out from under the tail. Your kid's first poop will be sticky, black, dark meconium that may have a green, brown or yellow tint. Meconium can appear as a chain, a sticky rope or a softer stool. As your kid's diet changes from colostrum to milk, its feces will be softer and more yellow, and may dirty the hair around the tail. The doe will clean this area of the kid. Watch for signs of diarrhea. Watery stools that stain the legs and leave the back end of your kid wet is reason to call your vet.

Kids can be let out to run free in and out of the heated area during the day. A dry environment with no drafts is essential. As kids become proficient at drinking milk, they can tolerate the cold more and benefit from a greater "play" area that's not enclosed under the heat lamp.

You'll notice that, when they're tired, kids will return to the heat lamp to sleep. Watch your kids closely. If kids are piling up under the lamp, it means there's not enough heated space for them. If they're lying near the outside radius of the heat lamp's radiant area, then the lamp is too hot and should be raised. Kids are excellent jumpers and climbers. Your kid goats will try to jump and climb on heat lamps, so take care to attach the lamps securely with a metal chain. Try to place the lamps high enough so that kids can't reach them easily. Also keep in mind that big goats like the heat, too. You may have to create a "creep" area for the heat lamp zone that has an entry point big enough for only the small goats to get in.

When the weather is cold, goats cuddle up for warmth. Kids will snuggle with each other or their mothers or other older goats. A kid pile-up can be disastrous, with one poor kid suffocated at the bottom. Bigger goats sometimes smother kid goats by lying on them or squashing them between goat and wall. The more space you give your goats, the less chance of a kid being squashed and suffocated.

Kids love to jump up on things. If your water trough is a jumpable height for your kids, there's a good chance they'll try to jump on top of the trough, fall in the water and drown. Water buckets hung on the wall are a good alternative.

Kids can take summer heat quite well. On our farm, the warmest weather we see is around 25°C to 30°C (77°F to 86°F). If kids are too hot, they'll pant like a dog, with their mouth open. They can also be listless or abnormally lazy. Ears feel very warm to the touch and are not held in up, but rather flop or hang downward. If you suspect a kid is stressed from the heat, remove it from its current location and find a cooler one. In the summer, a shady spot on cool, damp grass with a breeze is ideal.

Dehydration is a concern when the weather is hot. Always have fresh, clean water available that your kids can reach even if they are on milk. When the weather is warm, a big concern is air quality. The warmer the weather, the faster bedding, manure and urine break down—creating toxic gases your kid will inhale. These gases, mainly ammonia, can cause pneumonia. Be wary of bedding conditions under heat lamps. While the barn and even the kid pen may smell clean, get down on your hands and knees and put your face where your kids are breathing air in the area under the lamp. If the air quality is offensive to you, it will also be offensive—and potentially harmful—to your kids.

Remember, clean, clean, clean is the key to healthy kid goats. The best recipe for kid survival is a clean environment with dry bedding, along with good nutrition and draft-free, good air quality.

Take time to observe your kids. Watch how they hold their ears and their position while sleeping, and check how much time they spend playing versus resting and sleeping. The faster you pick up on a subtle cue that your kids are not doing well, the better your chance of correcting the problem and avoiding a sick kid.

Work closely with your local feed store to provide the best nutrition possible. Feed may seem an extra cost but, if goats are fed properly, the investment will be easily returned through the productivity and health of your goats.

Feeding Time—Lending a Helping Hand

Sometimes kids have no interest in searching for a teat. Perhaps they have searched and searched and had no luck finding what they were looking for. Looking for an udder can be tiring and perhaps the kid has used up its energy stores. If a kid has not latched on after 20 to 30 minutes after birth (the colder the kid is, the less time you should wait), you should step in to help the situation along. If your doe won't stand for you, get someone to hold her while you tie the goat up or put her on a milking stand. I don't like to see a doe walking too much right after kidding, so try to keep her as quiet as possible. Grab the hungry kid and kneel down at your doe's side, facing backwards. The goat's teat will most likely have a "plug" in the teat end that prevents dirt from entering the udder and causing an infection. You will probably have to remove this plug if another, more assertive kid hasn't already managed to dislodge it.

Take the teat in your hand and squeeze it from top to bottom until the plug slips out. The plug is usually hard and a white to yellow color. It may be small—just a cap on the end of the teat—or it can be quite large, extending up into the teat canal. Once the plug has been removed, continue squeezing until you see a dribble of a milk-like substance. This is the colostrum, and you want to save as much as you can for the kids. Use one hand to guide the kid's head and mouth and the other to hold the teat. It may take some practice, and you may need someone to hold a teat for you while you guide the kid's head. Once the kid feels the teat in its mouth, the sucking reflex should start. But if it doesn't, encourage the kid to start sucking: While the kid's mouth is around the teat, try squirting a small amount of colostrum from the doe's teat into the kid's mouth.

Sometimes a doe will not stand long enough for a kid to latch on and start to drink, or she may kick at her kid. But once you get a kid latched on and sucking, the doe typically relaxes and lets her kid drink. A doe's udder can be very sensitive or sore and the first suck from a kid can be painful. Once the milk starts flowing, the pressure on the udder is relieved and the goat feels much more comfortable. The sucking also stimulates the doe's oxytocin production. This hormone makes the doe feel good and causes her to bond with her kid and whatever other animal is around, including you. Goats can be very affectionate during this time.

Bottle-Feeding Kids

If, however, you've tried and tried and your kid still will not drink, you'll have to carefully milk the doe by hand and bottle-feed the colostrum to the kid. To bottle-feed your kids, you'll need:

- A minimum of two 500 mL (1 pint) bottles or one 1 liter/quart bottle
- Two rubber kid or lamb nipples to fit bottles
- Funnel
- Clean, fresh colostrum for newborns less than 24 hours old; or milk or milk replacer (mix replacer according to instructions on the bag)
- Small pail with warm water
- 10 mL syringe
- A comfortable spot (warm, if in winter) to sit

The quality of milk that you feed to kids must be of good quality. Raw milk that you milk from your goat will stay fresh for 24 hours if refrigerated. At room temperature or warmer, milk spoils very quickly. When feeding kids, do not warm up more milk than you think they'll drink. (Don't keep the leftover milk that your kid doesn't drink for the next feed. Give it to the dog or the barn cats instead. They'll love it.) Always feed your kid fresh, clean milk in clean bottles.

Bottle-feeding kids is a job many human kids love. My oldest son loved bottle-feeding so much that water was put into the bottle instead of milk to avoid overfeeding the kids!

Depending on its size, a newborn kid will drink between 250 mL and 500 mL at the first feeding. Very small kids may drink a bit less.

Many milk replacers are on the market, some better than others. Good milk replacers can be mixed until they're completely lump-free; inferior ones can't. Lumps can cause digestive upset that leads to bloat and potential death. Milk replacers for kid goats can be either cow milk–based or goat milk–based. At our farm, we feed over 100 kids every year on a cow-based milk replacer. Over 12 years, we've raised more than 1,200 kids on a cow-milk-based milk replacer without any problems. Difficulties arise only with poor-quality milk replacers. If you end up with a milk replacer that doesn't mix well, put it in a blender or use an immersion blender to mix it.

Milk replacers tend to come in large bags with a relatively high price tag. The one we buy is $100 a bag and yields approximately 250 quarts/liters of milk. That means I can feed my kid goats for 40 cents a liter/quart, which is much cheaper than feeding them goat milk. When you have a lot of kids to feed, milk replacer works well—we go through at least 20 bags in a season. But if you have only two kids to feed, one bag of milk replacer will easily feed them until they're weaned. It may be worth the investment if

Tips— Bottles and Nipples

- **Just about any bottle will work for feeding kids.** I prefer 500 mL (1 pint) plastic bottles that are easy to hold. You can use glass bottles but they can break and you'll then need to carefully clean out all bedding that may contain broken glass.

- **There are a few different styles of nipples suitable for feeding kids.** All have pros and cons. Some are direct flow-through, which means the milk, especially thick colostrum, flows easily through them. The con is that the colostrum or milk may flow faster than the kids can swallow—they easily choke or aspirate. Some nipples have a closed nipple tip with an X cut in the top. When the kid is sucking, the X opens and allows milk to flow through, but otherwise the milk does not drip out. The problem with this style is that kids must have a strong suck to get milk out—they may not work well for newborns.

- **Your best bet is a baby bottle made for human babies.** It's plastic, doesn't usually leak when inverted, and the milk comes out readily with a gentle suck. It's also easy to get milk out of the bottle by giving the bottle a gentle shake. The only drawback is that, as the kids grow and their suck becomes stronger, they can actually suck the nipple right off the bottle.

you want to reserve the goat milk for your family. If you have only one kid, some of the bag of milk replacer may go to waste unless you feed the rest to another animal.

The process

Bottle-feeding kids is enjoyable if you have patience and the proper setup. It can take roughly 30 minutes to an hour to feed a kid with a weak suck or poor suck reflex, so allow lots of time. Never feed cold milk to a cold kid, which shouldn't have to waste its energy warming up the milk it is consuming. In a tiny body, cold milk can also lower body temperature to a dangerously low level. I like to use two bottles of milk so one of the bottles is always warm. Milk cools quickly in a cold environment, so if your kid takes a long time to drink, you can switch to a warm bottle when one gets cold. Place the used cold bottle back in warm water to reheat.

Remember that all kids 24 hours old and younger must receive colostrum. After the first 24 hours of life, milk or milk replacer—either cow or goat—is just fine.

First, wash your hands and make sure all your bottles and equipment are clean. Fill clean bottles with colostrum, milk or milk replacer. You may need to use the funnel for this job. Once bottles are full, attach

This position works well for a kid that is familiar with a bottle. For newborns, try holding them on your lap so they can't pull away and cradle their heads while introducing the nipple.

nipples and place them in the pail of warm water. The milk should be slightly warmer than a goat's body temperature (39°C/ 103°F). The water in the pail should be slightly warmer than the milk. If the milk in the bottles is cold, use hot water in the pail. The milk should feel pleasantly warm to the sensitive skin on your wrist. If the milk feels too hot, the kid's throat will scald.

Once the milk bottles are ready to go, place the pail near your seat, ideally where the kids and goats cannot bother you. Goats are notorious for "helping" and making a big mess. Either that or they will irritate you so much by climbing on you, jumping on your back or head, nibbling on you or rubbing on you that you'll quickly lose patience feeding your kid. Find a place where other goats will not disturb you.

Get your kid and sit down. I'm most comfortable sitting on a milk crate or a small square bale of hay or straw (a chair is too high off the ground). It's best for your kid to stand while drinking—that position allows for the proper flow of milk through the kid's digestive tract.

If your kid does not yet know how to suck from a bottle or if it is too weak to stand, place the kid on your lap with its legs curled under so it's in an upright position. You do not want your kid's head on the same level as

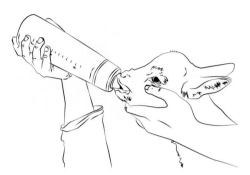

Support the kid's head while introducing the nipple. Kids are not born knowing the bottle is a source of milk. Many resist the bottle at first, but they soon learn what the nipple is for!

A kid pen with a chicken feeder used as a feeder for goat starter.

its stomach. Support the kid in one hand and hold the milk bottle in the other. It may feel awkward at first but, after a few tries, you'll get the hang of it. I like to support the kid's body with my left arm, lap and body, while I hold and support the kid's head with my left hand. I place my hand under the kid's jaw—thumb on one cheek, and fingers on the other (see illustration). In this position, I can guide the nipple into the kid's mouth, hold the nipple in place and stimulate the kid's suck reflex. I can also feel when the kid swallows.

Before you introduce the bottle to your kid, make sure the nipple is working properly. When you tip the bottle upside down, the milk should drip out, not flow. If the milk does not drip, gently squeeze the nipple. If the milk drips easily, proceed. If it takes a bit of work to get the milk out, carefully cut the hole of the nipple to make it slightly larger. If the milk flows out quickly, do not tip the bottle up too far or you will allow more milk into the kid's mouth than it can swallow. This causes milk to go into the trachea and possibly into the lungs. If a kid gets even a drop of milk in its lungs, it is almost certain to develop pneumonia—and death is a probable outcome.

With your right hand, hold the bottle comfortably upside down. Guide the nipple into your kid's mouth. You may have to squeeze its mouth open with your left hand. Once the nipple is in your kid's mouth, it may start sucking and keep sucking. If that happens, your job will be straightforward. If it doesn't, you have a bit of work in front of you.

The Sucking Reflex

Depending on the strength and health of the kid, however, activating the kid's sucking reflex may require different strategies. For some kids, the desire to drink milk is enough to start them sucking. You may see a kid sucking on its own tongue or hear sucking sounds as the kid searches under its doe for a teat. With other kids, it's not that simple.

Stimulations

Here is a list of increasing stimulations meant to activate the kid's suck reflex. Start at the top of the list and work your way down.

If nothing on this list works, sometimes placing your kid somewhere warm for 20 minutes or so will help. I've found that if they are even slightly cold, scared or uncomfortable kids will not drink, no matter what. It sometimes helps to step back a bit, placing your kid on the floor to walk and relax. After you've both had a break, you can start again.

Place nipple or teat in the mouth. First place your finger rather than the nipple in the kid's mouth, gently stroke the roof of the mouth (as that may stimulate your kid to suck) and then introduce the nipple or teat once sucking starts. Introduce a slight in-and-out or side-to-side movement of the nipple in the mouth.

Withdraw the nipple and re-introduce it to the kid's mouth. Place a drop or two of milk on the kid's tongue (if you have the right type of nipple, gently shake the bottle while the nipple is in the kid's mouth). You may end up with a quantity of milk running down the outside of the kid's mouth. If that happens, you're being a little too aggressive with the milk. Try tilting the bottle down a bit to lessen the flow.

Check the positioning of the kid in your lap. Make sure it is truly comfortable. Some kids will suck only in the presence of their doe, so give that a try as well.

If you have no luck with any of the above, a small quantity of milk in the kid's stomach sometimes wakes a kid up enough to activate the suck reflex. In this case, you're also trying to activate the swallow reflex by effectively force-feeding the kid. If you're successful, the presence of milk in the belly may increase the sensitivity of the suck reflex.

To feed a kid this way, fill the syringe with colostrum and hold the kid in your lap as described for bottle-feeding, with the kid's body and head supported with your left arm and hand and your right hand on the syringe

Tips—Creep Areas and Feeding

- **If your kids are running with the does, you will have to make a creep area for them.** A creep area is an area into which the kids can "escape," but where the bigger goats can't fit. If your barn is like mine, there are many points where the kid goats can find a place of their own where the adult goats can't fit, but the little ones can fit in easily. If this is the case in your barn, simply set up a kid station with starter ration, a nice warm bed (add a heat lamp if it's very cold), hay and water. The kids will quickly find and value this space, though you may have to guide some of them to it. Prepare the creep area by the time the kids are at least two weeks old.

- **If your pens don't have spaces that are small enough for kids to squeeze through, build a pen within a pen.** The walls of the smaller pen should have an opening big enough for kids but not so big that an adult goat can gain access. You may have to outsmart your mature goats in this situation. Most goats will not put up the effort to get into the kids' creep area, but every once in a while you'll have a doe that will do just about anything to get at the kids' grain or to lie under a heat lamp, if there is one.

- **A feeding guideline at our farm is one to 1.5 liters/quarts of milk per kid per day, depending on body type** (stocky, more heavily muscled kids need more milk), divided by the number of feedings in a day. For our own sanity, we feed all kids twice a day (even the newborns), and I recommend you do the same. The only time we do more than two feedings is when the kid is not very strong and hasn't been drinking enough milk at one feed or when the weather is extremely cold (less than −20°C).

- **Having fresh, clean grain in front of kids at all times is critical**—kids may gorge themselves on grain and bloat if they have been without grain for some time. Start with a sprinkling in the bottom of a pan or feeder. The kids may not eat much the first few days, but they will explore. Clean the sprinkling of grain out twice daily at chore time. As the days pass, the kids will need increasing amounts to be satisfied. Eventually, you will notice the grain is gone. The kids have eaten it all. At that point, gradually increase the amount of grain given. The trick is to strike a balance between giving the kids enough to keep them from getting hungry and not so much that they are wasting grain.

A doe doing a good job looking after her kid.

(switch hands if you're left-handed). Place the end of the syringe into the left side of the kid's mouth about halfway down the length of the tongue. Aim the syringe slightly downward so the milk runs down the tongue to the back of the throat. Do not squirt milk directly into the back of the kid's throat. You should be able to feel your kid swallow. A swallow means milk has reached the belly. Keep your bottle nearby, and if the kid starts sucking on the syringe, quickly substitute the nipple for the syringe.

If force-feeding doesn't do the trick, place your kid somewhere warm, under a heat lamp, and try again in 20 minutes or so.

It is critical to get a minimum of 500 mL (1 pint) of colostrum into your kid in the first 24 hours after birth. More and sooner, of course, is better.

Remember that you cannot feed your kids too much. If the milk/colostrum is of good quality, meaning the milk is clean and the milk replacer is lump-free, your kids will be fine. We always feed our kids until they can't possibly drink anymore. That doesn't mean you should force your kids to suck after they have stopped drinking on their own. It means allowing your kid to quit drinking on its terms, not yours. When kids are small, your goal is to get 500 mL to 1 L of milk into your kid in a 24-hour period. If your kid drinks 250 mL in the morning and the same amount at night, that's

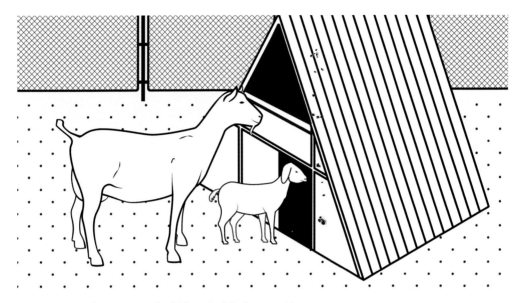

An example of creep area for kids suitable for outside.

fine—feeding twice a day is all you need. If your kid drinks only half that amount, you'll have to increase the number of feedings over the course of a day.

Once you get to know your kid, you'll learn exactly how much milk to feed it; its belly will feel full and round afterward. Check the soft area of the abdomen just behind the rib cage. Feel this area before and after feeding. You will very quickly learn the difference between a full stomach and an empty one.

After the first 24 to 48 hours, your kid should have the hang of drinking milk. After the first few days, kids will know the routine of when they get fed and will demand milk from either you or the doe, depending on whether they are being bottle-fed or not. During the first week or so of life, kids will not consume anything but milk. After this initial period, they will start to explore their options and start nibbling on grains and grasses or hay and straw and anything else they can get their mouths on. It's best to have free-choice access to a grain starter ration and good-quality fine hay or grass. If your kids are to be on pasture with your adult goats, worms will most likely be a problem. Talk with your vet on how and when you should de-worm your kids.

An adult goat's milk production is directly related to how that goat was raised as a kid—proper feeding and housing is critical to producing good milking goats. A starter ration for kid goats is specially formulated to have the proper amount of energy, protein, vitamins and minerals that your kid needs to grow up into a beautiful, healthy, strong goat. It's high in fiber and formulated with what are called buffers. Buffers, such as sodium bicarbonate (baking soda), are present to protect the kid's stomach if it consumes too much grain. If a ruminant consumes too much grain, the rumen's pH can become acidic, which causes the rumen to shut down. When it stops working, the goat stops digesting. The buffers in the grain ration help prevent the stomach environment from becoming too acidic.

This high-fiber diet is also important to the health and development of the kid's rumen. If you can't find a starter ration specifically formulated for goats, one for calves is the next best thing. We've raised hundreds of kids on a calf starter ration but I have had better results with the kid ration. Some starter rations come medicated. Your kids do not need a medicated ration. If you do have a problem with your kids, talk to your vet.

If your kids are being bottle-fed and are kept separate from your mature goats, it's easy to provide a pan of fresh, clean starter ration and a dish of water in their pen. Kid goats, even though they are on a fluid diet of milk, still need free-choice access to fresh, clean water.

Kid Housing

Healthy kids need a clean, dry environment with somewhere warm to rest—sleep is a must. Clean, dry bedding and fresh, clean water (kids need water even when on milk) are necessities. Be sure to clean wet areas daily and add fresh, dry bedding as necessary. If the weather is cold—less than 10°C (50°F)—provide a heat lamp or warming mat for the kids. We have used warming mats meant for piglets with great success. If the weather is very cold—less than –15°C (5°F)—provide a box for the kids. A simple homemade plywood box or a large plastic barrel with a hole cut in the end as a door and a heat source inside will accommodate a few kids. The best thing about the plastic is that it is washable.

Feeding kids whatever ration the doe is eating is fine. If you have the choice between a textured feed and a pellet, choose the pellet. Textured feeds look delicious, but the pellets ensure a balanced diet because they don't allow the kids to pick through and just eat what they like.

If you're using a water trough, do not fill it with water that's deeper than a standing kid is tall. Kids have been known to drown in a water trough. If you have a tall water trough and imagine that the kids can't jump that high, guess again. Kids are fantastic jumpers. They also love climbing, so a tall water trough may be seen as a direct challenge to their skills. The result of achieving this goal, however, is often not a successful one. A drowned kid is all too often the result.

Weaning

Basically, weaning is the process of ending a kid's access to milk. Once a kid is six to eight weeks old, it should be chewing its cud regularly, a sure sign that its rumen is functioning. The rumen is one of the four chambers of a ruminant's stomach. Located just behind the rib cage, it supports a healthy population of bacteria that consume the kid's diet of starter ration and hay. A thriving rumen is also a sign that the nutrition the kid receives from milk is no longer necessary. If your kid is in good health—no colds, no diarrhea, no running nose—it's time to wean it.

If your kid is bottle-fed, weaning is easy. Simply stop bottle-feeding the kid. If your kid is drinking milk from her doe, separate the kid and doe. If a kid does manage to get to its doe's udder after being off milk for a period, it can gorge itself—and a large amount of milk added to the digestive system at this stage, especially sporadically, upsets the population of necessary bacteria in the kid goat's rumen. That in turn may cause bloat. If kids are weaned at the right time, they will not look back. If kids are weaned too early, before the rumen is properly developed, the kid will be more susceptible to illness.

If you do not see any cud chewing, or if you are unsure, wait another week. If you regularly make detailed notes describing your kid's behavior, your observations may help you make a difficult decision.

If a kid has ready access to her doe, it may also wind up eating another animal's grain ration. The same applies to the doe. A doe may not normally seek to get out of her pen but, if she's trying to reach her kid, she may do just about anything to get out. Once she's out, she can also get into feedstuffs that are not meant for her.

Weaning a doe from her kid is not always traumatic. Depending on their personality and disposition, some does may attempt to wean their kids by headbutting or running away from the kids when they try to drink.

Once your kid has been successfully weaned and is thriving without milk, gradually switch from a starter ration to a developer ration. Feed a mixture of both rations for three to four days to ensure an easy switch and allow for gut flora to adjust. You may also find a regular milking ration works well for your kids. Remember to always follow recommended feeding rates.

Continue the starter ration (or other recommended grain ration) diet until the kids are of breeding age, following the instructions on the feed tag to determine amounts. Once these kids reach breeding age (around seven months or 32 kg / 70 pounds), they should be beautiful, sleek, fit and able to successfully carry a pregnancy, deliver strong, healthy kids and produce an abundance of milk. Kids raised on hay and insufficient grain, however, will have rough, scraggly hair coats and potty bellies. They may have difficulty becoming bred and maintaining pregnancy. If they do deliver, chances are they will have a difficult birth and weak kids.

Clean feeders, clean grain, clean hay, clean water, clean air and clean, dry pens are all keys to raising happy, healthy kids.

CHAPTER 5

Milking Your Goats

THE NUMBER ONE rule when milking an animal, when the milk is intended for human consumption, is that everything must be kept clean. Anything that touches the milk must be washed in warm water mixed with an alkaline-based detergent, then sanitized in warm water with a chlorine (or other) based sanitizer. Hands, buckets and milking units (if you're using them) must all be clean.

Why do we need to be worried about cleanliness? It's all about bacteria. There are useful bacteria, and then there are bacteria that can wreak havoc on the flavor and consistency of cheese. But there are also pathogenic bacteria to worry about. These are bacteria that make the people who eat your cheese sick. They include, but are not limited to, *Listeria*, *E. coli*, botulism, *Salmonella*, *Brucellosis* and *Cryptosporidia*. A bucket left out in the open air in your barn for a few hours can pick up any number of nasty bacteria. If left in an unclean pail, these bacteria will grow.

Because cleanliness is the goal, the best material for milking equipment is stainless steel. Repeated cleanings and scrubbings do not diminish its quality. If it develops rust or other stains, you can bring it back to shiny new with elbow grease. Stainless steel equipment may be a little pricey, but it holds its value—you can use the equipment for years and still sell it for the price for which you bought it.

Plastic, in contrast, is almost impossible to clean. It scratches easily, and these scratches are difficult to clean. Microbes love to live in the shelter of a scratch. In their world, that scratch is like a giant valley, a perfect place to hide from the cleaners and scrub brushes of our world. Do *not* use plastic pails for your milk collection.

In this book, I focus on hand milking because it is the most affordable way to get milk from your goat, and if you have only one or two goats to milk, it isn't a time-consuming chore. Your arms and hands may get sore at first, but after milking a few times, your muscles will adjust to their new daily task.

However, if the task of hand milking becomes a strain, you may wish to invest in a small unit to milk your goats. A number of small hobby units, both new and used, are on the market, and the seller should be able to show you how the unit works.

To Milk a Goat

Place all your milking equipment in a sanitizing solution of chlorine and *warm* water. Allow these items to soak while you fetch your goat.

Don't use hot water, it deactivates chlorine. I don't recommend using bleach to sanitize your dairy equipment. Bleach has added fragrances that will taint the flavor of the milk. Buy a proper dairy sanitizer and follow the instructions for mixing a sanitizing solution that's food grade and does not need to be rinsed.

Bring your goat up to the milking stand. Some people feed their goat a small amount of grain during milking. I don't. Goats are easy to train and don't need an incentive to stand and be milked. They enjoy the procedure and the attention.

Wash your hands and collect all your milking supplies from the sanitized water. Place all the equipment within easy reach, where it can't be contaminated.

Clean your goat's teats with the udder wipes or udder wash and towels.

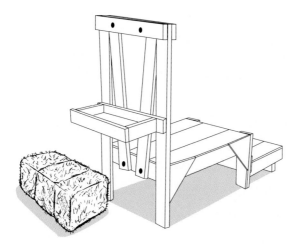

Example of a goat milking stand.

Getting Started

You will need:

- **A source of hot and cold running water** near the milking area.

- **A large wash sink.** A big laundry-tub-style sink will do, but the ideal is a stainless steel dairy sink deep enough to hold all your milking paraphernalia. Bigger is always better.

- **A stainless steel pail with a lid.** Size is determined on how many goats you're milking and how much milk they produce. A 10 L (2.6-gallon) pail is a standard size and should easily hold all the milk two goats produce on any given day.

- **A small stainless steel container.** Goats are quite talented. It is amazing how fast they can pick up a foot and put it down right in the middle of the beautiful, hard-earned milk you have just collected. So, rather than losing the entire pail of milk with one kick, I fill a small container and then pour it into the big pail. The smaller container should feel comfortable in your hand and fit underneath the goat as you're milking it—a 1L (4-cup) container is fine.

- **Hand soap and clean towels** for washing and drying hands

- **Udder wipes or udder wash and disposable paper towels**

- **Teat dip**

- **A milk filter holder.** Filter as you go to save time and eliminate a milk-handling step. Adapt the filter mechanism from a large retired coffee pot so it can sit atop the large milk container. There are also filters made specifically for straining milk—a pricier option to discuss with your dairy equipment supply company.

- **Milk filter.** Milk filter size should match your milk filter holder and can be purchased at your local food or dairy supply store.

- **A goat stand.** You'll be milking once or twice a day, every day, so it's well worth the investment of time and money to create a comfortable place for milking—the less strain on your body, the better. A variety of small stands, both new and used, are made specifically for milking goats. You can also make your own. If you don't mind being on your knees, tying your goat works as well.

- **A goat collar.** Some goats stand very nicely to be milked without being tied. Some don't. Experience will teach you what works for you and your goats.

Washing and dipping

There are two important reasons to wash udders before milking and dip to protect udders after milking.

First, you are producing a food product that's meant for human consumption. Your goat's udder sees its fair share of manure and urine in a day. Just because your goat's udder looks clean doesn't mean it is.

Second, you put a lot of time, energy, love and money into your goats. It would be a shame to lose a goat or goat milk from mastitis, an infection of the udder that can damage it beyond repair. Teat dips not only contain an anti-microbial agent (usually iodine) but also skin emollients and conditioners that will prevent cracks and other sores on your goat's teats.

Udder prep and post-milking dipping are two small things you can do for your goats' well-being—and your own. They are simple, quick and relatively inexpensive.

The milking process

After your goat's udder and teats are clean, it's time to milk your goat.

Start by placing the milk container under your goat's udder. Experiment with positioning to find what works best for both you and your goat. You may want to sit in front of the goat's hind leg or behind it. Once you're comfortable, it's time to get your first precious amount of milk. If you've never milked before, it may be a challenge at first. But once you understand the mechanics of milking and get the

Udder Wash versus Udder Wipes

The initial cost of udder wash makes it more expensive to use than udder wipes, but if you're only milking a goat or two, it costs less than a penny per application. It's much like washing with soap and water. Rub or dip the wet, soapy solution on the teats and then rub it off (taking the dirt with it) using paper towels. Using an inexpensive teat dipper, a container made specifically for this purpose, is the easiest way to apply the udder wash.

Instead of using paper towels, you can use cloth dairy towels that can be laundered. The towels must be properly washed with a sanitizing agent to prevent the spread of potentially harmful bacteria from one milking to the next. A dairy supply company is the best place to purchase these towels, which are made of a material that's gentle on your goat's udder and is able to absorb moisture and remove dirt and debris.

Udder wipes are quick, convenient and easy to use. They are like human baby wipes but are bigger, stronger and contain more antimicrobial properties. Simply pull a wipe from the container and rub your goat's teats until they are clean.

This goat is comfortable with Katie milking her and does not need to be tied.

feel of it, it will be a quick and easy job.

Start at the top of your goat's teat, right near the udder, and squeeze the milk from top to bottom. You do not want to pull on your goat's teat. It is a squeezing motion from top to bottom. You're squeezing the milk from the top of the teat out through the teat opening at the bottom, much like squeezing icing out of a pastry bag. Frequently empty your small milk container, through your milk filter, into the big container holding all your milk.

I like to switch from one teat to the other, switching hands to keep my arms and hands from getting too tired. This process also allows time for the milk to drop down into one teat as you're milking the other. You can tell that your goat's udder is empty when the milk flow slows and the udder feels flat. When full of milk, an udder feels much like a water balloon. When empty, the udder moves much more freely under your touch.

All goats' udders are different. Some appear to be quite large before milking and quite small after milking. Others don't change much in size or shape. There's a wide difference in the shape and size of your goat's teats as well. Sometimes teats are so small that you can barely get a thumb and forefinger on to milk. Others are wide and long, almost touching the ground. Few udders are perfect (though goats with excellent ones do well at any goat show). All udders and teats come with

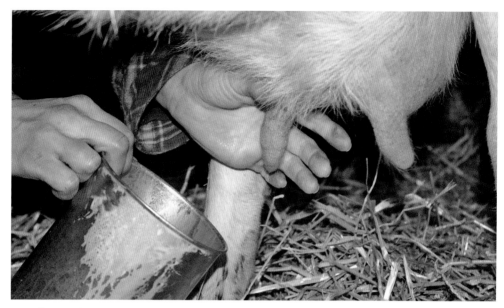

Start by gently but firmly grasping the top of the teat.

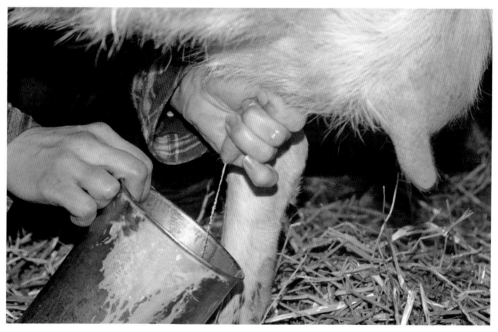

Squeeze the teat from top to bottom forcing out a stream of milk.

Milking Your Goats

Release the teat and start again. Repeat on both teats until both sides of the udder are empty.

a unique set of potential problems and benefits. Some are easy to milk; others are not. Some are prone to infection; others are not. Some teats are easy for kids to find and latch on to; others are not. Over time, you will become an expert on goats' udders and how they work.

After your goat is milked, dip teats with an iodine dip that's specially formulated for them. The teat end remains open for about 20 minutes after milking. The teat dip will prevent any bacteria from entering the udder via the teat and thus helps prevent infection.

Questions about milking

How much milk does my family need?
Depending on the breed, genetics, stage of lactation and management, a single dairy goat can produce between 1 and 6 liters/quarts of milk a day. More is possible, but it's not that common. If your family of four requires 2.5 liters/quarts of milk a day, one goat should be enough. If your family needs less milk, you will have plenty left over with which to make yogurt and cheese.

Raising and Keeping Dairy Goats

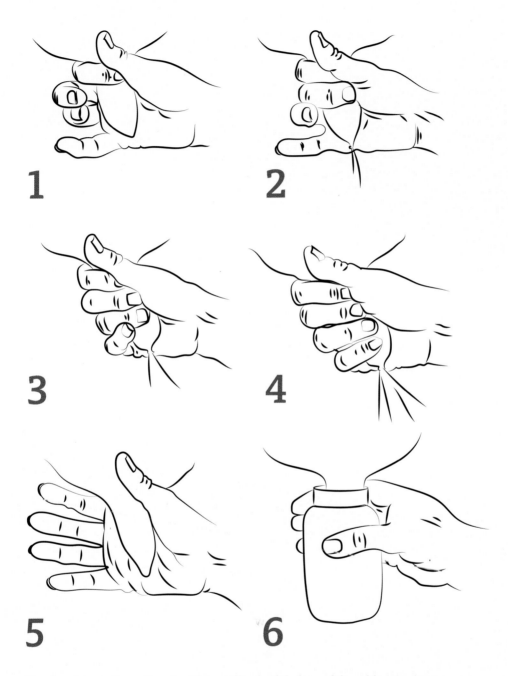

Step-by-step motion of hand milking. After both halves of the udder are empty, dip teat with an antiseptic teat dipping solution.

Can I freeze milk?

If you have a surplus, you can freeze goat milk after it has been pasteurized. Bacteria present in pasteurized milk will continue to multiply slowly at freezing temperatures, however, and milk can go sour even if frozen. For best results, use the cleanest containers and equipment possible and put milk into a deep freezer as quickly as you can after pasteurization and cooling. Cool the milk first before freezing it.

If milk is left refrigerated for a few days before freezing and the cream separates and rises to the top of the milk, the cream will freeze separately and look like small grains of rice when you thaw the milk later. The milk will taste fine, but the texture isn't the nicest. The small clumps of fat in the milk will melt when it reaches body temperature, so warming the milk slightly before consuming it helps fix this texture problem. You can make cheese from frozen milk, but neither the yield nor the quality will be as good as if the milk were fresh. Soft fresh cheeses such as chèvre freeze well. Yogurt does not.

Can I milk our goat if she's still feeding kids?

If you want your doe to nurse her kids and would like to take milk for yourself as well, you must milk her at least once a day from the time she kids. Twice a day is better—the more you milk your goat, the more she will produce. Before you milk, make sure the kid(s) have been up to nurse and have full bellies. The rest of the milk is then yours.

As the days pass, the amount of leftover milk should increase if your doe is comfortable and well fed. As the kids get bigger and start nibbling at solid feed, you can take your portion of milk first and leave the leftovers for the kids. Your doe will continue to produce milk for the kids during the day and night.

Common sense is your ally. If you have a high-producing doe and there's always lots of milk for you after the kids have had their share, and if the kids are big, strong and healthy, you can safely take a greater share of the milk before your kids drink. If your doe does not have much milk left over after the kids have had their fill and the kids are a little on the thin side, you should continue to let the kids drink first. If your doe does not have any milk left after the kids have fed and the kids are struggling to grow, you must concentrate on your doe's well-being. Wait until the kids are weaned before taking any milk.

Proper nutrition, good health and a good environment are the keys for a doe to produce lots of high-quality milk.

Can I miss a milking?

Once you start on a regular milking routine, you cannot miss a milking. Your goat produces milk according to what is being taken from her. If she's used to being milked morning and night, and one night she does not get milked, she still produces the same amount of milk. Skipping a milking not only eventually results in a decrease in milk production but can also lead to mastitis. Skipping a milking is also very uncomfortable and painful for your goat, especially if she's in peak milk. If you're going to be away for a milking, arrange for someone to milk in your absence.

While you shouldn't miss a milking, it is possible to change your milking schedule to suit your needs. For example, if you normally milk at 6 a.m. and 5 p.m., and you have a 6 p.m. dinner date, milk at 3 p.m. instead of 5 p.m. and again at 5 a.m. instead of 6 a.m. Resume your regular schedule the following night. If you are late (or early) for a milking by an hour or so, you won't upset your goat too much.

What is a goat's milk production cycle?

A goat's milk production cycle is determined by when she becomes bred and delivers her kid(s). Before the goat delivers her kid, the udder will prepare itself for milk production. It will start to produce and store milk, a process that's often called "bagging up." Do not milk your goat before she kids, even if her udder looks very large. Milking your doe before she kids can cause mastitis, premature labor or a decrease in the available antibodies in the colostrum that are vital to the kid's health. Once the kid or kids have been delivered, your goat will be ready for milk to be taken by either a kid's mouth or your hands. With each milking of the goat, the colostrum gradually changes into transition milk.

The general rule is six milkings in total, in other words, twice a day for three days, before your goat's udder produces actual milk. Let your kids have all the transition milk. If you want your goat to produce lots of milk, it is critical that you hand-milk her twice a day during this period. If you're worried the kids are not getting enough milk, bottle-feed this milk back to them. As long as she's being milked, a goat will produce milk for eight months to a year or more. The volume of a goat's milk typically peaks at eight to 12 weeks post-kidding.

A goat's milk production also greatly depends on day length and weather. A goat naturally produces the most milk during the warm summer months, when the days are long. As the days begin to get shorter, her milk production begins to drop. The warmer and drier the fall is, however, the higher a goat's milk production. If the fall is cold and wet, a goat's milk production drops rapidly and will most likely stop altogether once the snow flies. This drop in production is also determined by other factors, such as breed and individual genetics, nutrition and a goat's level of health and comfort.

How can I increase our goat's milk production?

The best way to encourage a goat to produce as much milk as possible is to make her comfortable and keep her in good health. Take time every day to observe your goat and her surroundings. The most critical factor in good milk production is a goat's environment. A goat needs lots of warm, dry, clean bedding on which to rest and sleep. Despite having the best feed and an ample supply of fresh water, if a

Synchronization

It's best to synchronize your goat's breeding and kidding so she will be in milk during the warm summer months when fresh grass and food are plentiful. You may wish to breed for off-season milk if you want milk in the winter, but it's not easy to do because goats are typically seasonal producers.

goat has nowhere to lie comfortably to rest and produce milk, she will produce very little.

When she is not asleep, a goat in milk should be eating, drinking, playing, walking or resting. If your goat is just standing around doing nothing, she needs attention, and it's up to you to figure out what kind. Does she have enough bedding? Is it clean? Is there good air quality in her resting area? Ammonia levels are strongest closest to the ground. The air may smell clean where you're standing, but that doesn't mean your goat is standing in a clean area. Does she have easy access to fresh, clean water and feed? Is another goat picking on her? Are her kids missing? Go into her space and figure out what's wrong. If you can't find any problems with her environment, take a close look at the goat to see if there is anything wrong with her.

I've milked our goat and have fresh milk. What's next?

Make your milk as cold as possible as quickly as possible. A refrigeration unit is best. Be sure to agitate your milk often until it is completely cool. A second option is a cooler with ice, especially necessary in the warm summer months. You should have no problem keeping milk cold in the winter when the outdoor temperature is below 4°C (39°F). You can also pasteurize your milk right away and heating warm milk to pasteurization temperature will take less time.

What is good-quality milk?

Good milk means clean milk. Clean milk means a low bacteria population, and that population should be free of coliforms and *E. coli*. Coliforms present in milk are a sign of poor hygiene. *E. coli* present in milk means the milk has been contaminated from manure. Dirty udders and teats, unwashed hands and dirty, unsanitized milking equipment are all likely causes of coliform and *E. coli* contamination. A warm teat filled with milk is the perfect place to incubate *E. coli*. Udder and teat washing and teat dipping after milking are critical.

It's also possible for milk quality to deteriorate when it is stored after milking. A container that has not been cleaned and sanitized before use adds undesirable microorganisms to your milk. One of the most annoying, although not harmful, bacteria is *Pseudomonas*—cold-tolerant bacteria that thrive in refrigerated milk. They create a soapy taste in milk and cheese and appear as a fluorescent pink, green or yellow blemish on your cheese. These bacteria will also glow under a black light. After all your hard work raising a goat, milking her and making the cheese, it is a great disappointment when potentially good cheese is blemished because of improperly cleaned equipment.

It's true that plenty of good bacteria are naturally present in milk. Unfortunately, unwanted bacteria have the potential to make you seriously ill—another reason to be highly disciplined about hygiene as you milk.

Even when milk is as clean as possible, enzymes and bacteria are still naturally present and will affect its quality. Milk above 4°C (39°F) deteriorates quickly. Cool milk to between 1° and 4°C (34° and 39°F) and maintain this temperature until you're ready to use it.

If your raw milk is to be used for fluid milk, it should be less than 24 hours old. When I'm making cheese, I don't like to use milk that's any older than three days. But if it is good quality milk with low bacteria count to begin with *and* has been kept very cold, raw milk can be used up to five to seven days after milking.

Keeping your milk outdoors during the winter keeps it cold, but do not place it in the sun. The sun not only warms your milk but also oxidizes it, which produces off flavors.

Milk quality also depends on the amount of milk solids present in the milk. Goat milk is about 87 percent water, 4.5 percent lactose, 3 to 6 percent butterfat, 2.8 to 4 percent protein and 0.7 to 1 percent minerals and vitamins. There can be a great difference in fat and protein content in the milk of each goat, and these differences are determined by the goat's genes, stage of lactation and nutrition, as well as by the time of year. By comparing tastes and cheese yields, over time you will be able to tell when the milk's fat and protein content is up or down.

Raw Milk versus Pasteurized

In Canada, it is illegal to sell or give raw milk to anyone but immediate family members who are living under your roof. Some U.S. states allow raw milk sales, but these sales must come from registered and inspected farms—you cannot sell raw milk simply because you have it. Remember that it is not illegal to produce and consume raw milk from your own animals, but it is illegal to give away or sell raw milk without the proper approvals. (It is also illegal to sell pasteurized milk unless you own a registered and inspected dairy facility.)

Raw milk can be safe. You can drink it, and many people do, and never get sick. You can also drink raw milk and become deathly ill. Most of us have strong enough immune systems to fight off any pathogenic bacteria that may be present in milk, but small children, the immune-compromised and the elderly are highly susceptible to becoming ill from the bacteria that may be present in raw milk. Of special concern are pregnant women. The bacteria *Listeria*, which is commonly found in raw milk, soft fresh cheeses and even in aged raw milk cheeses, can breach the placenta wall and affect an unborn child. According to the Centers for Disease Control and Prevention (CDC), that can lead to premature deliveries, miscarriages, stillbirths and serious health problems for the newborn.

Milk that is contaminated with pathogenic bacteria does not look odd or smell or taste off. There is no way to tell if milk contains pathogenic bacteria other than to have it tested at a lab or to consume it and see whether you get sick. Contaminated raw milk can come from the cleanest-looking barns and the healthiest-looking animals, which is why, in my view, drinking it is like playing Russian roulette.

As farmers, we know there are certain diseases that pass from doe to kid, cow to calf or ewe to lamb through the milk. The first milk a kid, lamb or calf receives is actually not milk but colostrum. Colostrum plays a vital role in

Pasteurization

Heating milk to a known temperature for a known time with the purpose of killing pathogenic organisms is pasteurization. There are legal pasteurization temperatures that are recognized by governments to improve milk safety to the general public. The legal pasteurization temperatures and times are too hot for pasteurizing colostrum. Heating colostrum to 60°C (140°F) for 60 minutes is certainly "pasteurization," even if it is not legal for sale. Care must be taken to not heat colostrum above that temperature. It will turn into useless pudding if made too hot.

the immunity of the newborn animal. Feeding good-quality colostrum to a kid can be the difference between life and death. Without it, the fate of a young animal is most likely death within three weeks of birth. But we also know that colostrum, like milk, has the potential to pass along disease to the newborn. For this reason, we pasteurize colostrum at 60°C (140°F) for one hour before feeding it to kids. All the antibodies that a kid needs for a thriving immune system are present in the pasteurized colostrum. The only thing the kid does not get from the pasteurized colostrum is the disease-causing bacteria and viruses that we don't want passed from doe to kid.

Both the U.S. and Canadian governments allow the sale of raw milk cheeses that are aged more than 60 days from licensed dairies. This 60-day rule works well for hard cheeses with low moisture content, because bacteria such as *Listeria* need moisture to thrive. While the drier cheeses do not provide a great environment for *Listeria*, they are a great home for some of the good bacteria that live in cheese. Over the 60 days, the good bacteria outcompete the *Listeria*, which is trying to survive in less than favorable growing conditions. However, it is also true that *Listeria* has been found in hard cheeses older than 60 days.

The raw milk 60-day rule can be deadly for soft, fresh cheeses such as quark, caeso blanco, chèvre, cream-style cheeses, cottage cheese, Brie and Camembert. These cheeses have a high-moisture, low-salt content, and cheeses like Camembert and Brie have a pH that's very close to neutral. All these conditions create the perfect place to grow the bacteria that are going to make you deathly ill. Nor does refrigeration lessen the danger. Cold temperatures may slow bacterial growth, but they don't prevent bacteria from growing. And bacteria such as *Listeria* and *Pseudomonas* actually thrive in the cold, damp temperatures found in these fresh cheeses.

In short, fresh cheeses can be more dangerous to the consumer than raw milk, which is why I strongly recommend that you always pasteurize the milk you use to make fresh cheeses.

Even if your fresh cheeses are consumed immediately after they are made, they can be extremely dangerous to eat. *E.coli* present in fresh cheese can cause deadly disease. I know one cheese maker who ended up in the hospital and almost lost a kidney from eating fresh cheese that had been made from raw milk. In my opinion, it is not worth the risk. The home pasteurization procedure I outline will give you milk that's very close to the nutrition of raw milk without the dangers of consuming raw, or uncooked, milk.

At our dairy, we have monitoring equipment, fancy vats and mounds of paperwork that ensure our milk is properly pasteurized. But the process our milk goes through is no different from how you (or I) might pasteurize milk in our own kitchens—a process that differs from the high-temperature, short-time pasteurization used by big commercial milk processors.

I choose to keep my family as safe as possible while providing the best nutrition possible, which is why I milk goats and pasteurize my milk. I also choose to keep our customers as healthy as possible by selling them the best-quality, safest milk and dairy products possible.

Home Pasteurization

There are many home pasteurizers on the market. You may wish to invest in one, but there's a good chance you already own the necessary equipment.

The goal of pasteurization is to heat milk to a temperature of 62°C (144°F) and hold it at this temperature for 30 minutes. Care must be taken to not overheat the milk or pasteurize for longer than necessary.

Materials
- A stainless steel double boiler that's big enough to hold all your milk. If you can't find a double boiler, two pots, one bigger than the other, will be fine. One pot holds your milk and nests inside the second pot. The second pot holds water and acts as a water jacket. The milk pot must be stainless steel. The pot that's the water jacket can be non-stainless, although stainless steel is best. (When you're making pasteurized

Two thermometers are useful when pasteurizing milk. One thermometer measures the milk, while the other measures the water bath temperature.

cheeses, this milk pot should double as your cheese vat.)
- A journal and pen
- A tight-fitting lid, to cover all the pots
- A consistent heat source
- A large stainless steel spoon
- A dairy thermometer that clips onto the side of the pot
- A timing device
- Containers for milk storage
- A funnel (if necessary)
- A large sink or bucketful of ice water to quickly chill your milk
- Sanitizing solution, detergent and a source of hot water to wash up

Process
1. Wash your hands.
2. Make sure your work surfaces are clean and free of clutter.
3. Sanitize your work surfaces and any equipment that will touch the milk. Sanitizer must be mixed as directed by the manufacturer's instructions. A dairy sanitizer is best, without the perfumes that chlorine bleach has. When mixed properly, sanitizer solutions are food grade and do not need to be rinsed off.
4. Record the date and amount of milk you are pasteurizing. You may wish to include other details, such as the name of the doe that produced the milk and what type of feed your goat has been eating.
5. Fill water-jacket pot so that it will not overflow when the second pot is nested into it.
6. Place water-jacket pot on the stove over high heat.
7. Pour cold raw milk into milk pot. (If you are pasteurizing milk immediately after milking, there is no need to chill milk first.)
8. Nest milk pot carefully inside water pot, stirring milk constantly.
9. Monitor temperature of water in water jacket. Agitate water periodically to ensure even heating. The milk will not get warmer than the temperature of the water. Turn heat source off when water temperature reaches 64°C (147°F).
10. After turning off your heat source, monitor milk temperature while constantly stirring. The target temperature of the milk is 62°C (144°F).

11. Once target temperature is reached, turn off heat, stir milk to ensure even heat and then put the lid on your pot and hold it at 62°C (144°F) for 30 minutes. Record the time. Prepare ice water bath to cool your milk once the 30 minutes are up.

Milk and water have a great capacity to hold heat. If your work area is relatively warm (20°C/68°F or warmer) and draft-free and you have a tight-fitting lid, your milk should hold this temperature.

Once you've pasteurized milk a few times, you can confidently leave your milk unattended for 30 minutes. But until you're comfortable with all stages, check the milk temperature periodically by removing the lid and stirring your milk as you take the temperature. Remember, each time you remove the lid, you risk losing heat and may have to briefly turn on the burner again.

The same property that ensures heat retention in water and milk also means it takes a long time to raise the temperature of your milk. It is wise to heat slowly and not to allow your water jacket to get too hot—it's very difficult to stop or slow the heating of the milk once you note the temperature is getting too high. Don't worry if the milk reaches a few degrees above pasteurization temperature. That will not ruin your milk. However, if at any time your milk falls below 62°C (144°F), your milk is not pasteurized. Milk is pasteurized only if it stays above that temperature for 30 minutes.

Record the stop time on your pasteurization. Write down how many minutes the milk was held at pasteurization temperature. Record any problems or efficiencies.

Once the milk is pasteurized, remove it from its water bath, remove the lid and immerse the milk pot in ice water. Stir your milk and monitor the temperature until the milk has cooled. If you're going to make cheese right away, cool the milk to the desired temperature for adding cheese culture. If you're keeping it as fluid milk to drink, once the milk has cooled significantly (10°C/50°F or colder is best), fill sanitized containers with your beautiful pasteurized milk. Place bottled milk in the refrigerator and keep at 1°C to 4°C (34°F to 39°F) for the longest shelf life.

If your sanitation practices are good and you keep your milk very cold, your milk can last as long as three weeks.

Reminder: Even though your milk is pasteurized, milk is legal for sale only from a facility that is licensed by your local government. You cannot sell or even give away any milk you produce.

CHAPTER 6

Caring for Horns and Hooves

DISBUDDING AND HOOF trimming are procedures that either you or your vet can take on. Whether you're willing to take on these tasks is up to you, but it's helpful to know what's involved.

Disbudding

Every once in a while, you will see what is known as a "polled" goat, a goat born without horns, thanks to a recessive gene that must be present in both parents. But for the most part, both male and female goats have horns, and at some point you must decide whether you want your newborn goats to have horns. If you want to disbud your goat, the procedure is recommended to take place when the kid is between seven and 10 days old.

There are several reasons there's a preference for a goat without horns.

At our farm, the biggest argument for disbudding involves safety—for your family, your staff and your other goats. If you handle your goat regularly, there's a good chance you might be accidentally "stabbed" by a goat horn. A goat may simply move her head the wrong way and cause an injury. I've had fingers badly pinched when taking goats with horns out of our milking stand. Fingers can also easily be broken in the same way. These injuries can happen when a goat is not being aggressive. (Our goats have been aggressive only with other goats, never with their human companions.)

But the damage a goat with horns can do to another goat far exceeds the damage a hornless goat can do. We once had a young goat with horns that completely shredded the abdomen of a pen mate. These goats were getting along just fine—there had been no signs of aggression in the pen. One morning, I came out to find a half-dead, blood-covered goat lying on the ground and another very smug-looking goat with blood on her horns. It was not a happy moment for any of us. With some tender, loving care, the victim of the attack survived. We removed the horns of the offending goat.

A goat with horns can also get caught in the most unusual places—from gates, trees and fences to other goats' collars and head gates. If you find a trapped goat soon enough, the crisis is easily resolved. However, a goat in this situation can also quickly die. The simple stress caused by being caught and a lack of water or feed can all cause death. While monogastrics (animals having a stomach with a single compartment) like humans can live for a time without food, ruminants—especially goats—do not fare so well.

A horn is a living appendage. Unlike deer antlers, which fall off once a year after the blood supply to the veins and arteries is cut off, a goat's horns are permanently attached, and the nerves and blood vessels running through them are active. Losing a horn is comparable to losing an appendage. Sometimes when a goat is caught by her horn, the horn breaks. If a horn does not come off cleanly, a goat can bleed to death. A torn artery spurts blood everywhere. Sometimes broken horns require cauterizing, which is a much more traumatic experience for a goat than the process of disbudding.

All this is not to say that horns have no use. Horns make great "handles" for catching and guiding a goat. Our bucks are trained quite well to move by horn leading. Horns are temperature regulators in hot climates because increased blood flow through the horns helps to cool the goat's body. Horns are also a great defense mechanism. If you're planning to keep your goats outside on pasture or in the woods, where predators such as coyotes are around, a goat can defend herself quite well if she has horns. Indeed, she won't have much chance without them. Horns are also beautiful, and they can be quite majestic on a full-grown goat. The decision to keep or remove the horns is yours.

The goal of disbudding is to sufficiently damage the horn cells so they do not grow. This task is accomplished by cauterizing the horn tissue. That sounds and looks worse than it really is.

Keep in mind that the window for this procedure is very narrow. Horn buds must be visible but not too big to cauterize properly. A special hot iron has been designed for this purpose. It looks like a big soldering iron with a cylindrical steel tube on the end.

Hoof Trimming

If your goats spend a lot of time outside and are walking, running and climbing over rough or rocky ground, their hooves need to be trimmed less often than those of goats that spend much of their lives indoors on soft bedding.

Trimming hooves is relatively easy. As with any task, practice makes perfect and, as always, the proper tools make all the difference.

A variety of hoof trimmers are available, but my personal favorites are ARS shears. These trimmers are small and fit comfortably in my hand. Other hoof trimmers open too wide, and my grip is just not big enough to pull them closed again. But people with large hands find these trimmers easy to use as well. They are sharp and compact and make easy work of the hooves.

How to do it

To trim your goat's hooves, you may first have to catch and restrain the animal, though if your goat is used to being handled, she may stand for you without being tied or held. If you need to restrain her, tie her in the pen or in another spot that's easy to work in, one with ample lighting. For this job, I put my goats in the milking stand, since standing there is part of their daily routine and an environment in which they are relaxed.

Start with the front feet—these are the easiest to hold. It's a good way to introduce your goat to what you're doing. Remember, you're just trimming their hooves, not chopping off a limb. Trimming hooves is a lot like cutting your fingernails—it doesn't hurt, but it may feel odd. Some goats may think the procedure is horrible. Others don't mind in the least.

Understand your goal before you begin. Keep in mind the perfect shape of a newborn's or young kid's hooves. If you don't have a newborn goat for reference, look at the photos and illustrations in this section. Basically, you want to achieve a flat-surfaced foot that's parallel to the hairline where skin meets hoof. The flat surface is the

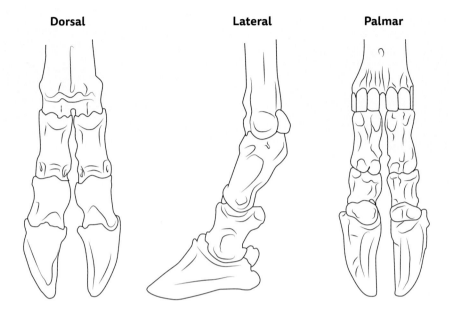

The bones of a goat's foot and lower leg.

surface that comes in contact with the ground. You don't want your goat to have a lopsided gait because one side of her hoof is longer than the other. If the hoof is parallel to the hairline, there should be balance between length of toe and heel.

Pick up the hoof and hold it in a position that's comfortable for both you and your goat—she'll let you know if it isn't. Take the closed foot shears and use the pointed end to scrape any dirt from your goat's hoof. If your goat's foot has been neglected for some time (or if she has fast-growing feet), there may be a flap of hoof growing over the foot. Remove it by cutting from heel to toe. Alternatively, if the growth is very long, cut off the extra-long growth first and then work from heel to toe. Do not cut the hoof from toe to heel, which leaves the sharp, pointed end of the shears facing you. If the goat moves quickly, which goats are prone to do, you can be easily cut by the trimmers. Yes, that's happened to me.

When you reach fresh new tissue, stop cutting. Fresh tissue is clean looking and a whitish color. If you inadvertently hit blood, don't panic. Let the hoof bleed to clean out infection. It may look like a lot of blood, but it will clot and stop. Keep a close eye on your

Caring for Horns and Hooves

One possible position for trimming feet.

goat over the next few weeks in case the wound becomes infected and an abscess develops. If that happens, you may notice your goat limping, and the hoof will feel warm to the touch. In my experience, these abscesses clear up nicely—on their own—in a few days. However, if your goat shows lameness for longer than three or four days, call your vet.

Once you've trimmed away the bulk of the hoof, move on to creating the "perfect hoof" according to the picture in your mind. As you look at the hoof, you will notice a "wall" around the outside and a softer inside sole. The wall of hard hoof tissue is your chief target—first one side, and then the other. You will naturally cut back the sole as you trim. As you cut, turn the hoof and compare the angle of the foot to the hairline.

Remember, the bottom of the hoof should be as parallel as possible to the line where hoof meets skin. At the same time, check the flatness of the sole of the foot. The inside of the hoof should be on the same plane as the outside is. The inside of the foot can be cut slightly shorter because it does not wear as quickly as the outside, but the outside should never be shorter than the inside.

You may not get the hoof perfect on your first trim. That's okay. It's better not to have a perfect foot than to cut the hoof too short. Get back at it in a month or so. Make small changes over time. If you've acquired an older goat that has had improper hoof care or none at all, the foot may be badly distorted. When this happens, the inner live tissue grows deeper into the horn, or dead tissue, of the hoof. It's very difficult to trim a foot like this. You may hit the soft tissue and draw blood easily and unexpectedly. Again, don't panic. Finish trimming the hoof as best you can while being extra cautious. Take small amounts off. And remember, it's better to stop too early than to take too much off. You can always go back and continue later, after the hoof is allowed to grow.

Once the front hoof is finished, proceed to trim the rest of the feet, using the same approach. Hind feet are harder to hold, so if your goat is

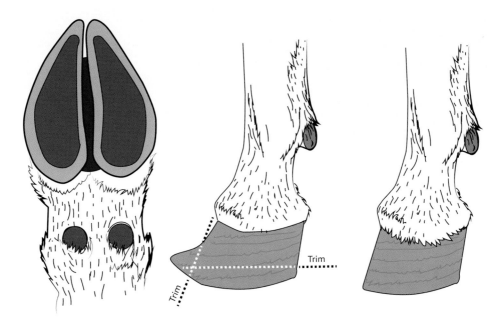

A view of the bottom of a goat's hoof, before and after a trim.

literally kicking up a fuss, ask someone to help hold her leg. You may be in for a tug and pull. Goats can be persistent, but you can be even more stubborn. Don't give up. Tell your goat that getting her hooves trimmed is for her own good, and keep going. It gets easier and easier every time you trim. If you become too tired, put your goat away so both of you can rest and relax, and try again later or the next day.

As with all procedures, keep careful notes. If a hoof was cut too short, make a note of which one, along with the goat identification and date. Record any other relevant details. No matter how good your memory is, it's never as good as your written word. Next time you trim hooves, read your notes first to remind yourself of problem areas.

A Goat's Hoof

Your goat's hoof is made up of hard, dead tissue with soft, live tissue underneath. It's just like a human fingernail. If you cut off too much of the hard tissue and bite into that soft tissue, the hoof will bleed. To avoid bleeding, remove small pieces of hoof at a time. Remember—you can always cut more hoof, but you can't put it back.

Caring for Horns and Hooves

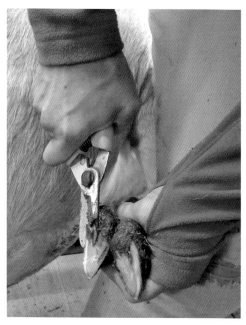

Trim small pieces of hoof away at a time to avoid removing too much hoof and nicking the live tissue. Always point trimmers downwards and away from your hands to avoid nicking yourself!

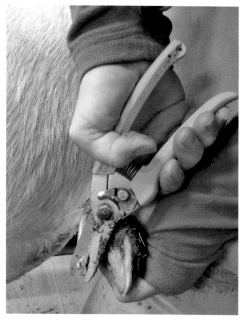

Keep trimmers as flat as possible in relation to the bottom of the goat's foot.

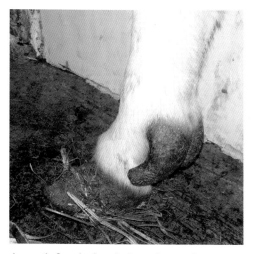

A goat's foot before being trimmed.

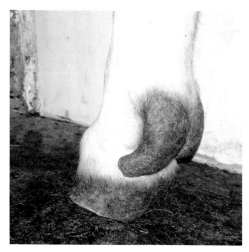

A goat's foot after being trimmed.

CHAPTER 7

Disease and Illness

THERE'S AN OLD saying that "a sick goat is a dead goat." Sadly, there's some truth to it. Goats that become ill rarely make a full recovery, and often, though not always, a goat dies as a result of an illness. If the goat survives, it's often no longer strong enough to be a productive asset, which is especially relevant in the case of a milk-producing animal.

A number of diseases are common to all goats, but some cause great harm specifically to the dairy goat. Since the practical value of a dairy goat to her owner is her milk, diseases that affect milk production are of particular importance.

One of the best ways to keep your goats healthy is to manage your herd well and provide dry housing, clean bedding, ample space with a source of fresh, clean water and abundant high-quality feeds. All these things are important to goat health. Air quality is also critical. Goats are extremely susceptible to respiratory illnesses such as pneumonia. The quality of the air is always poorer closer to the ground—make a habit of regularly getting down on your hands and knees to be sure your goats have what they need. If you don't like breathing the air, your goats won't either. While you're down there, take note of your hands and knees. If they are dirty and wet, your goats are also dirty and wet. Many diseases and illnesses stem from dirty, wet environments with poor air quality.

An equally valuable tool in your goat first-aid kit is simply to know your goats as well as you can. At our farm, this is the first job we give to new staff: find a comfortable seat, somewhere inconspicuous, and sit and watch the goat herd. If a person isn't familiar with what constitutes normal behavior for a particular goat, it's impossible to pick up on the subtleties in behavior that a goat exhibits when she's ill or on the way to becoming ill.

At the top of your to-do list, add an annual visit from your veterinarian. During this visit, ask any and all of the questions you may have about your goats.

Review the diseases outlined in this section. Ask which drugs, vaccines and de-wormers the vet recommends for common ailments. Take detailed notes. That way, if a problem does arise, you'll be prepared to deal with it, and you may just manage to avoid an expensive after-hours vet call. (In my experience, an emergency call to the vet rarely happens during regular working hours.)

In your search for a veterinarian, you will probably learn that not many vets are goat specialists and even fewer are experts in the intricacies of the dairy goat. That said, there might be a vet in your area with a special interest in small ruminants. If there is, you're in luck, since much of the science that applies to a sheep also applies to a goat. The next best option is a vet who specializes in dairy cows, since much of that science can likewise be transferred to goats. Keep in mind that if there's a vet in your area who has emigrated from another country to North America, he or she will probably know more about goats than does a native vet, simply because goats are valued far more in other countries than they are here.

Given this gap in veterinary medicine, it is even more important that you do everything you can to prevent illness and disease. The bottom line is that no matter what your vet's strengths are, the relationship has to work for both of you. Sometimes a vet who is honest enough to admit a

Goat Facts

- **Age of puberty:** anywhere from four to 12 months, depending on the goat's nutrition and growth rate.
- **Length of estrous cycle:** 21 days (average), range: 18 to 23 days.
- **Length of heat period:** 18 to 24 hours is average, but the range is 12 to 36 hours. Standing heat (the time during which the goat is willing to stand to be bred) can be a short as six to 12 hours.
- **Gestation:** 148 to 153 days; average is 150 days.
- **Birth weight:** 0.9 to 4.5 kg (2 to 10 pounds).

lack of knowledge about goats but is passionate and committed to finding a solution to the problem can be your most valuable asset.

As my aunt always said, "Don't give me the best doctor—give me the second-best doctor, because they'll actually have the time to care for me."

Temperature

If a goat is approaching the extremes of the ranges described below, keep a careful eye on her. If she is well outside the range, there's an obvious problem that must be addressed. Remember, these are simply guidelines.

A healthy dairy goat's body temperature typically ranges roughly between 39°C and 40°C (102.2°F and 104°F). A high temperature indicates that the goat has a fever, and a fever is in turn typically caused by an infection. Fever can also cause a goat to become dehydrated.

Low temperature indicates hypothermia (the core body temperature drops, and vital signs grow weak); if a goat is dehydrated and in warm environment, it may be close to death. A low temperature may also suggest that a goat is experiencing problems digesting.

I rarely take a goat's temperature with a thermometer. Once you know your goats, you'll be able to tell by observation or touch when one has a fever or a low body temperature.

It's just like being a parent. Ears are a wonderful indicator of fever. If your goat's ears are warm, it has a fever; likewise, if it's "tucked up" and shivering or has droopy ears. (If you have LaMancha or LaMancha crosses, the ear droop rule doesn't apply.)

That said, a rectal livestock thermometer is a lifesaver when you're dealing with a hypothermic kid. A kid can be laid across your lap and the thermometer lubricated for easy insertion with human saliva or a little bit of petroleum jelly.

When your goat is chilled, its ears, other extremities and the inside of its mouth are cold. Hypothermia is rarely seen unless the goat is very young or wet and in a cold environment. The coldness is most often caused by dehydration.

A dairy goat's heart rate is 70 to 95 beats per minute. Stress from illness or pain can cause an increased heart rate. Lack of energy, dehydration or cold indicates a lower heart rate. In this scenario, typically, a goat is close to death.

A dairy goat's rate of respiration is 10 to 30 breaths per minute, while the rumen movement is 1 to 1.5 per minute. Increased respiration suggests the goat may be suffering from pneumonia. It may also be reacting to poor air exchange in its dwelling. Decreased respiration can be a sign that not enough energy/oxygen is going to the goat's body. The goat could soon enter a coma. Note that goats pant quickly like a dog when they are hot.

Rumen Movement

Rumen movement is a little harder to detect than heart rate, temperature or respiration. If a goat is stressed, it may stop chewing its cud, which makes it difficult to assess rumen movement. Once at ease again, the goat resumes chewing. With this in mind, you can better assess your goat's comfort level when it encounters a new situation, such as a change in housing, being milked for the first time, being held for an examination at the vet's or being transported to a new location for a show or fair.

Checking whether your goat is chewing its cud is the simplest way to assess if it's ruminating properly, but it's also possible to note rumen movement by observing the side of a goat, behind the ribs. This movement is especially visible when the goat is lying down and the rumen (or big belly, in human terms) is pushed up. The rolling action of the muscle contraction of a rumen looks and feels very much like a baby's movement in a pregnant human being, so rumen movement is often mistaken for kid movement when a doe is heavily in kid or in the process of giving birth. Rumination, or digestive sounds, can also be heard with a stethoscope or even an ear pressed to the left side of a goat, again, in the area behind the ribs. Lots of sound is always a good sign. An absence of sound or movement can indicate a lack of rumination.

If your goat is not ruminating, it may be becoming acidotic (see Acidosis, below). Encourage the drinking of electrolytes and eating hay. A solution of water and sodium bicarbonate (baking soda) can also be offered, depending on the severity of the condition. I fill a 60 mL syringe and slowly squirt the fluid on the left side of the goat's mouth on the back of the tongue. This is the best way to prevent aspiration of the fluid. Hold the head up for the entire process, and wait to feel the goat swallow before releasing the head. (Electrolytes can also be given this way.) If your goat will ingest hay on its own, don't force anything.

In addition to these vital signs, behavior and production are other important signals of good goat health and well-being. However, it's not always the case that inconsistent behaviors indicate illness: they may simply indicate discomfort. Even so, by remedying the source of discomfort, you both increase the quality of life for your goats and decrease the incidence of illness and disease.

Worrisome Signs and Behaviors

Watch for the following indicators of possible health problems:

- **Amount of water consumed.** Changes in a goat's pattern of water consumption can be caused by environmental temperature and humidity, milk production, cleanliness and quality of water, feed changes or illness.
- **Feed intake.** Appetite can be influenced by water intake, quality of feed, stage of lactation, pregnancy or illness.
- **Manure and urine.** Changes can be caused by feed and water intake, change in feed and illness.
- **Milk production.** Increases or decreases in milk production can be caused by stage of lactation, length of daylight, feed or water intake, environmental temperature, goat comfort, stress or illness.
- **Changes in the udder or milk.** These changes can be caused by injury, infection, stage of lactation, pregnancy, disease and illness.
- **Changes in personality.** These changes can be caused by increases in discomfort or comfort, changes in social status, pregnancy, stage of lactation and illness.
- **Changes in body condition (fat and muscle).** These can occur when there are changes in feed intake, feed quality, stage of lactation, pregnancy, change of herd social structure, illness and disease. Always note incidents of lameness, bumps, swellings, cuts and bruises.
- **Changes in hair coat, skin and eyes.** Changes in these areas can be caused by parasites, feed quality, water intake, weather, environment and illness. Goats are very clean creatures. A dirty hair coat is a sign of a poor environment, an illness or discomfort. A goat's hair coat should be smooth, shiny and lying flat (though you should always allow for differences between hair

coats of different breeds). If a hair coat is standing up, the goat may be experiencing discomfort such as cold or pain or overall poor health. A rough hair coat is the result of poor nutrition. Flakes and dandruff can be caused by parasites such as lice, nutrition, habitat, illness and disease. Bacterial infections can cause a multitude of skin ailments. Cloudy eyes might be a sign of illness or indicate a problem with the eyes themselves. If the mucosal lining of the eye is pale in color, that could be an indicator of intestinal worms or illness.

- **Inaction.** Right now, what is your goat doing? As I remind everyone who works with us on our farm, every goat in a pen of goats should always be doing something, whether it's eating, drinking, scratching, playing, sleeping, resting, chewing or ruminating. If a goat is standing motionless in a pen, chances are there's something wrong. It may be something simple—perhaps it can't find a dry place to lie down. But inactivity may also indicate there is something more complex at play. It could be a sign of illness or injury; there might be a dominance problem in the herd. Use your detective skills to determine just what's bothering your goat. Throwing in some fresh bedding may do the trick, but sometimes more careful examination or observation is required.

Other causes of discomfort

Other causes of discomfort can include anything from weather to the goat's general environment. The goat may have a problem with anything from bedding to feed quantity and quality, air quality, lighting, noise, social interactions and stability of environment (goats don't like change). Sometimes, it can be the oddest thing that causes a personality change in your goat, and only the most careful observer is able to figure out the cause. Is your goat grinding its teeth? If so, you can be sure it's in pain.

Enjoy your goats. Spend time with them. To assess a goat's health, it's important to know and understand how a healthy goat looks and behaves, and that can be done only through hours of careful observation.

If your goat is sick

A designated sick pen located in a dry, well-lit, draft-free area of your barn is a valuable asset to herd health. Make sure the pen is pleasant and accessible for the care-giving humans, too. If it isn't, a sick goat may not get the attention it deserves. At the same time, the pen should be far enough away from the other goats that they are unable to touch noses with the patient. Ideally, they shouldn't even share a common wall.

A sick goat is not looked upon with sympathy by its fellow goats. It may even be pushed out of the herd, which makes it tough for the goat that's under the weather to get proper nutrition, water and a dry, warm place to lie down. The sick pen helps to contain disease while giving sick or failing goats a chance to rest and get proper nutrition without the stress of worrying about possible interference from more dominant goats. Put a sick goat in the pen or quarantine a new-to-the-herd goat for a few weeks, and be sure to clean and disinfect this pen in between occupants.

On our farm, we use a product called Virkon to sanitize our pens. It's a pink, multi-purpose, powdered disinfectant that's mixed with water according to the package directions. The resulting solution can be sprayed onto pen walls and floors with a hand-held sprayer (such as the ones you might use to apply lawn sprays). Virkon is more effective than bleach because it kills bacteria and viruses.

Common Diseases in Dairy Goats

Here is a review of some of the more common diseases found in dairy goats and a description of the impact they can have on both individual animals and herds. This alphabetized list is by no means exhaustive. A good reference is the *Merck Veterinary Manual*, which can be found online at merckvetmanual.com. Mary C. Smith's book, *Goat Medicine*, is my preferred source, but it is expensive and not easily accessible if you don't have an animal or a veterinary science background.

Please note that when a disease is described as "zoonotic," it means it can be spread from animals to humans. You should always use extra caution when dealing with animals that carry these kinds of diseases.

Caprine arthritis and encephalitis (CAE)

There is no cure for CAE—only treatment and prevention.

If you purchase goats, there's a good chance you'll be buying a goat infected with the CAE virus. A virus infection that's widespread in dairy goats across North America, CAE is considered by many to represent the greatest danger to dairy herds. As the name suggests, CAE presents as arthritis (swollen, disfigured, painful joints) or encephalitis (inflammation of the brain causing behavior abnormalities). Other symptoms can include an extremely hard udder with little milk produced after kidding, and lesions and abscesses on the body. Wasting and poor body condition can also be CAE symptoms.

A goat becomes infected with the CAE virus when a kid consumes colostrum from an infected doe shortly after it is born. The virus can also be spread through saliva (at feed bunkers and water troughs) and through blood (via the common use of needles).

Symptoms may show themselves at any stage of a goat's life, from the age of six months until death. Alternatively, they may never show themselves. This is one reason the disease is so prevalent. A doe can have a dozen or more offspring before any symptoms of CAE occur. Every kid that drank colostrum from the infected doe now carries the virus. Once these kids are old enough to be bred, they, too, will pass on the virus unless CAE prevention methods are used.

CAE prevention involves removing the kids from the does at birth and feeding them pasteurized colostrum, cow's colostrum or commercially prepared powdered colostrum. Goat colostrum can be pasteurized by following the method outlined in this book for pasteurization of milk (see pages 100–103). Decrease the temperature to 55°C to 60°C (131°F to 140°F) and increase the time to 60 minutes

During kidding season on our farm, we sometimes pasteurize colostrum several times daily. Three years ago, I discovered a wonderful new pasteurizing method that uses a slow cooker with a programmable thermometer probe. After putting my colostrum in good-quality zipper storage bags, I immerse the bags in a water bath in the slow cooker, programming it to shut off when the thermometer reaches 55°C (131°F). I leave the colostrum in the hot-water bath for 60 minutes, and the slow cooker holds the temperature constant. If you happen to have this style of slow cooker, by all means use it for this purpose. (If you have only one or two goats, it may not be worth investing in a new pot.)

When pasteurizing colostrum, stir often and don't overheat. Overheating causes the colostrum to quickly turn into a solid mass that's impossible to feed to kids. Heating it at too high a temperature may also inactivate the immune cells and antibodies that make colostrum precious. Without these antibodies, the kid is very likely to die from disease in the first few weeks.

Before feeding colostrum to a kid, be sure to cool to a lower temperature (around 40°C/104°F) to avoid scalding. Keeping a backup supply of powdered

A goat kid transportation option.

colostrum on hand is also highly recommended.

Caseous lymphadenitis (CL)

There is no cure for this zoonotic disease—only treatment and prevention.

People should always exercise caution when dealing with goats with CL.

CL is caused by *Corynebacterium pseudotuberculosis*, bacteria that can survive for long periods in the goats' environment. Antibiotics do not cure this disease.

Externally, CL manifests itself in large abscesses typically found on a goat's jaw under its ear; these abscesses can appear in other locations on the body, as well as internally. Visible abscesses should be lanced and drained and the wound flushed with an antiseptic. Before treating, always ask the advice of your veterinarian, and be sure to discuss a long-term health and treatment plan for your herd as well. Wear protective gloves, and take care not to spread the bacteria around the environment.

In addition to external abscesses, your goat may also have internal abscesses that you're unable to diagnose visually but which affect her organs. Abscesses on the lungs cause coughing, which in turn spreads more of the bacteria. Wasting—the chronic deterioration of an animal that shows itself in a loss of strength and muscle mass and a loss of appetite—is also a symptom of this disease.

On our farm, I have noticed an increase of external abscesses when a goat is under stress—for instance, when a doe is kidding. There is a vaccine that helps control the disease, which has resulted in a decrease in abscess frequency and size in our herd, but it will not prevent it. Once a goat is infected with CL, she's infected for the rest of her life.

Coccidiosis

This disease is treatable and preventable.

Coccidiosis is caused by a protozoa of the species *Eimeria* that is host specific, which means that coccidiosis in goats is caused by a different species of *Eimeria* than the one that causes the disease in sheep or in cattle. In other words, only goats can pass coccidiosis to other goats. The method of transmission is fecal to mouth. Young kids are extremely susceptible to the disease, first because their immune systems have not yet matured, and second because of a kid's insatiable curiosity and need to mouth everything.

The best way to prevent a coccidiosis outbreak in your herd is through cleanliness—access to clean, dry pens, ample space on pasture (or pasture rotation if goats are outside), clean water and feed are key to preventing this and many other illnesses. Over the 14 seasons we have had goats, we have had only one outbreak of coccidiosis.

With this disease, it is necessary to treat all goats that have had contact with the infected goat or shared common areas. Diarrhea and weakness are common symptoms in kid and adult goats. However, since diarrhea is a symptom of many different diseases and ailments, a confirmed diagnosis through a vet is necessary before treatment. Specific treatments for coccidiosis vary by geographic location, so please check with your vet before treating your goats.

Contagious ecthyma, or sore mouth—also known as orf

There is no treatment or cure for this disease—only prevention.

This zoonotic viral infection bears a close resemblance to warts. Starting out as blisters that become crusty scabs, contagious ecthyma most often manifests itself around the mouth of the goat, especially a kid—hence, the name "sore mouth." The wart-like lesions can also be found around the feet and around udders and teats if infected kids are nursing. Contagious ecthyma is caused by an extremely resilient poxvirus known as orf virus, which is capable of surviving in a barn environment for many years. The lesions themselves last for one to four weeks.

The good news is that once a goat has been infected with the virus, she builds a resistance to new infections. The bad news is that the virus is highly contagious and spreads rapidly through a herd. Quarantine of an infected goat is recommended. Ecthyma can also spread to humans, so be very careful when handling a goat with lesions. Wear protective gloves, especially if hand-milking a goat with lesions on her udder. If lesions are on the udder, take care to not pass the infection on to another goat with your hands or by udder washing, teat dipping or through the use of milking equipment.

Enterotoxemia, or overeating disease

Very few animals survive this disease so prevention is extremely important.

This disease seems to creep up on you when you least expect it. During evening chores on our farm, a group of yearlings may look fit, sleek and active. They are eating well and seem completely healthy. The next morning, one of the biggest and healthiest of these goats is lying on the ground, dead. There has been no warning.

Sometimes a goat lives a little longer; you might find one lying on the ground, weak and in pain. Sometimes a goat has diarrhea; other times, no symptoms at all.

Enterotoxemia is caused by *Clostridium perfringens*, bacteria that are present in every goat's normal gut flora. Usually, it isn't a problem, but if the perfect conditions occur, its numbers can explode, producing a tremendous amount of toxin that's lethal to the goat.

To prevent this disease, vaccinate kids with a CD-T vaccine as soon as possible after they reach the age of one month. This vaccine helps prevent *Clostridium perfringens* types C and D and *Clostridium tetani*, or tetanus/lockjaw. Enterotoxemia type C typically affects young kids; type D typically affects kids over one month old. Since enterotoxemia is common in young kids, it is also very important to vaccinate the bred does at least four to six weeks pre-kidding. The bred doe will pass her immunity on to her kids both *in utero* and through colostrum. This vaccine requires a booster according to the manufacturer's or veterinarian's instructions, typically a few weeks after the first shot, then annually, so keeping records of breeding dates and proper feeding of colostrum for your goats is very important.

Foot rot

Foot rot is treatable but very difficult to eradicate from a farm once it's present. Prevention is the key to avoiding this disease.

Our goats have never had foot rot, but it is quite common in goats. Since foot rot is caused by contagious bacteria that love warm, moist environments, keeping your goats in clean, dry bedding as well as giving them regular hoof trims are the keys to preventing the disease. Minimizing animal movement into your herd and asking visitors from other farms to wear protective plastic booties should also prevent foot rot, as well as many other illnesses and diseases, from infecting your herd. If you do suspect hoof rot, trim the goat's feet and douse with antibacterial solution. Wash and sanitize the hoof trimmers, clean out the barn and keep everything as dry as possible.

Lice

This condition is treatable but highly contagious.

Both sucking and biting lice can be found on goats. Lice are species specific. This means the lice that live on goats live only on goats and close relatives, such as sheep. A doe with lice will have a poor hair coat and an overall poor body language and appearance. She will also be itchy—she will scratch herself with her teeth as well as her horns and hooves if she has lice; she may scratch herself on gates and fence posts. The itchy goat should be inspected closely. Pull the hairs apart, and you will probably see small, dark, oval-shaped lice moving within the coat. Lice eggs are oval and cream-colored and attach to the hairs of the goat. Lice droppings are small black specks. If the goat has had lice for a few weeks, patches will appear in her coat and she will probably lose body condition. Make it a habit to check your goats' coats regularly.

If there is a goat with lice in your herd, you must treat the whole herd. Various treatments are available. If the goat is in milk, the product you use may contaminate the milk both internally and externally. Any drug given orally may be shed in the milk, while topical powders or drenches will inevitably end up on the goat's udder. Consult your veterinarian about recommended lice treatments.

To avoid an extra call fee, you might want to include this question on the list you keep for your annual vet visit.

Mastitis

Mastitis is inflammation of the udder; the most common cause is a bacterial infection. Symptoms can vary, but note that there may be no visible symptoms at all. Visible symptoms include off texture, color or consistency of milk; a hot, red, swollen udder; blood clots; and pus in the milk. Mastitis can also be caused by trauma to the udder itself. An aggressive blow to an udder by another goat can cause bruising that, if left untreated, can result in a bacterial infection.

If antibiotics are used in a commercial dairy goat, the milk cannot be sold, so in our herd, we try to avoid using antibiotics whenever we can. The milk from goats treated with antibiotics should not be consumed by the hobby-goat owner or given to other livestock or pets. Dispose of antibiotic-treated milk appropriately (recommendations on how to do so can be obtained from your local state or provincial agriculture department).

Although you cannot use milk from the infected half of the udder, you can legally use the milk from the healthy half of the udder as long as the goat has not received antibiotics. The decision to use antibiotics is sometimes a judgment call: there is a fine line between when antibiotics are essential in preventing the loss of the udder and when mastitis will clear up on its own without treatment with antibiotics.

Bacterial mastitis is typically a result of poor milking practices. Unclean hands, unclean udders and unclean or improperly working milking equipment can all cause mastitis. Once mastitis is detected, it is very important to strip the milk completely from the infected side of the udder at every milking or a minimum of twice a day.

The condition may clear up over time, but antibiotics may need to be administered. Many different bacteria strains can cause mastitis and different antibiotics treat them, so ask your veterinarian for advice on the appropriate antibiotic. A veterinarian can also advise on withdrawal times (when milk can be safely consumed) and udder management during treatment. Milking practices and preventive measures should also be assessed to avoid a recurrence of mastitis in the herd.

On our farm, mastitis cases are rare, thanks to proper goat care and management. In our experience, the most common cause of mastitis is an injury to the udder caused by another goat. Typically, we hand-milk and strip the infected side of the udder and dispose of the milk. A bruised udder causes a lot of bleeding, and blood will clot inside the udder and be difficult to remove. Massaging the udder helps to break the clots into pieces small enough to fit through the goat's teat. Whether a blood clot or milk clot, the clots *must* be removed from the udder regularly (twice daily) or a bacterial infection will result. Sometimes a great amount of force and a lot of manipulation is needed to remove the clot from the udder. Experience with hand-milking helps. Hand-milking and disposing of milk should continue

A mature milking goat enjoying a hug and a scratch from the author.

until the milk returns to normal. With regular milking, mastitis should begin to clear in 12 to 36 hours. If there is no improvement after this time period or if the mastitis appears significantly worse, it's best to consider using antibiotics.

Pink eye, or infectious keratoconjunctivitis

This condition is treatable.

Pink eye is typically caused by bacteria known as *Mycoplasma conjunctivae*, or by chlamydia infection. On our farm, the highest incidence of pink eye occurs during the warm, summer months when we have kids still on milk. On a farm, there are lots of places for flies to pick up bacteria on their feet. In addition, flies love milk, and they love poking around an animal's eyes, which is how they transfer this bacteria into the eyes of kids. Older goats can get pink eye, too, but kids are far more susceptible.

If you pay close attention to your goats, you may notice them displaying the first symptoms of pink eye—sensitivity to strong light or sunlight. A goat with an infected eye squints and appears to be in some discomfort. Usually the first observable sign of pink eye is a watery eye with pus. The pus might be dried out but can be extreme enough to seal the goat's eye shut, and it should be cleaned from the infected eye as soon as it is discovered. Soak a cloth in warm water and press it against the eye, softening the pus and allowing the eye to open.

Once the eye is clear of pus, apply an antibiotic ointment to the eye itself. We use an antibiotic mastitis treatment called Special Formula. We have never actually used this treatment for mastitis, but we always keep a tube on hand because it is the quickest, easiest and fastest cure for pink eye that I know of. Make a small line across the eye from corner to corner. The antibiotic should be spread on the actual eyeball and into the tissue surrounding the eyeball, as well as under the eyelid. The natural blinking action of the goat will help with the spreading around. An experienced goat handler can easily apply the antibiotic to the goat's eye, but it may be helpful to have an additional helper to hold the body of the goat while the antibiotic applier controls the goat's head. Obviously, kid goats are easier to restrain than adults.

The antibiotic usually clears up pink eye in one application, although occasionally it needs to be reapplied. Ask your vet for his or her recommended practice and product. Keratoconjunctivitis is a contagious disease, so pay close attention to all goats to make sure the infection doesn't spread. Treat all symptomatic eyes immediately. A failure to treat increases the rate of disease transmission and can also cause further complications such as blindness.

There are also non-infectious forms of this disease. Inflammation of the eye can be caused by hair, dust, dirt or hay. There is a congenital disorder (one that can be passed from parents to offspring) called entropion, which also causes pink eye-like symptoms. With this disorder, a kid is born with its eyelids turned in, which causes the eyelashes to rub against the eye and irritate it. With patience and persistence, the condition is treatable. Simply keep the eye clean with a diluted saline solution, and turn the eyelid out to its normal position a minimum of twice a day. To make a saline solution, mix one teaspoon of salt in one cup of warm or cool water. Always use a clean cloth and clean solution. I recommend using a single-use paper towel (soft white ones are much kinder than the harsher brown paper towels). Out of the hundreds of kids we have had born on our farm, only one has had this disorder. With treatment, she came along quite nicely and was a valuable asset to our herd.

Polioencephalomalacia (PEM)

This neurologic disease is treatable if diagnosed early enough.

Commonly referred to as "polio," PEM shouldn't be confused with the human poliovirus. These are two very different diseases; humans cannot contract the goat version.

The cause of PEM is thiamine deficiency. A healthy rumen produces thiamine, but if the bacteria population in the goat's stomach has been altered in some way or if rumen activity has slowed, polio may result. Its main symptoms appear to be those of an improperly working nervous system. Circling, stargazing (when a goat stares off into space), weaving, stumbling and blindness may be evident. These are also symptoms of other diseases, such as CAE, scrapie or rabies, but with PEM, treatment must be administered immediately to be effective. A thiamine injection is inexpensive and won't hurt the goat if it turns out not to be suffering from PEM. It's worth a try, especially if your goat has recently had a change in diet or may have overeaten a grain ration.

When we were new goat owners, one of our four goats escaped from her pen, found the grain bags and promptly gorged on grain. We assumed she was fine when she didn't experience any of the usual reactions to overeating, such as acidosis (a lack of rumen function that causes a drop in rumen and eventually blood pH) or bloat, within the first 24 hours. Within 36 to 48 hours after she'd overeaten, however, she began to exhibit some distinctly odd behavior: She was circling in her pen with her head arched backwards.

We had no idea what was wrong, so we called the vet, who eventually diagnosed rabies. The goat had to be euthanized, and since rabies is a reportable disease in Canada, the head had to be removed and sent to a lab for testing. Until the results came back (fortunately, they were negative for rabies), we were also legally obliged to quarantine our barn. Soon after that, we spoke to another vet who had a great deal more experience with goats. He told us that the simple cure for PEM is an injection of thiamine, or vitamin B1, with a recommended dosage of 20 mg thiamine/kg body weight.

This injection can be given either in the muscle (IM) or subcutaneous (SC)—under the skin. You can imagine our regret when we learned that a simple vitamin injection could have saved us from the death of our goat and the stress and worry of a possible rabies outbreak. A few weeks later, that goat's sister came down with PEM. This time, we knew what to do. After an injection of thiamine, she recovered almost instantly. Such is life with goats.

Pregnancy toxemia, or ketosis

This is a non-pathogenic disease (not caused by bacteria or virus) that is both preventable and treatable.

Pregnancy toxemia is a completely preventable condition that's common in poorly managed dairy-goat herds. The most critical time to manage a goat's nutrition is during the "transition phase," which begins three to four weeks before kidding, as a goat moves into late pregnancy, and lasts two to three weeks post-kidding, through the period when the goat progresses through birth and moves on to early lactation. During the first half of the transition phase, growing kids (especially multiples) dramatically increase draws on the doe's energy, protein and mineral reserves. At the same time, the growing kids inside the doe occupy much of the space needed by the rumen. This decrease in rumen capacity restricts the quantity of feed the doe can consume.

The act of kidding itself also is a huge draw on the doe's energy reserves. Milk production starts just before birth and increases quickly after kids are born. Nutrient requirements are high during this time. If energy, protein and mineral inputs from feed are not great enough, the goat pulls from her own body reserves to provide nutrients to growing kids or her milk supply. When too much of the goat's body fat is used for energy, the liver may be overworked and fail to process the fatty acids properly. The result is ketones that are toxic to the goat. Ketosis, or pregnancy toxemia, can be easily diagnosed via the sweet-smelling breath of the goat. Once you smell it, you will never forget the aroma. If the goat isn't treated, death will result in one to three days, sometimes even more quickly. Goats can be treated with propylene glycol, but if the goat is down and cannot get up, her chance of recovery is slim, especially if she has not yet kidded. Speak with your vet beforehand to have a good plan in place in case of pregnancy toxemia.

Goats that are most susceptible to

pregnancy toxemia are over-conditioned (too fat) with a body condition score of 3.5 or higher; goats that are carrying multiples; and goats that are undernourished during the transition phase. (See the Body Condition Chart on page 38.)

To prevent pregnancy toxemia:

Do not let does become overly fat during the last stages of lactation. As a doe's milk production decreases, so should her feed intake. Keep goats at a body condition score of between 2.5 and 3.5.

Goats need higher-quality forages as they enter the transition phase. Also, start feeding a grain ration to goats to ensure adequate energy intake. It is very important that you do not overfeed goats. Work with a livestock nutritionist or veterinarian to determine the right amounts of feed.

Have fresh, clean, ice-free water at the ready to encourage goats to stay hydrated.

Q fever

There is no easy, affordable cure or treatment for this zoonotic disease—only prevention.

Q fever is caused by the bacteria *Coxiella burnetii*, which can be carried by many different species, including humans. *C. burnetii* is most often seen in cattle, sheep and goats but can be carried by other animals, including dogs, cats, horses, pigs, rabbits, wildlife, rodents, birds and ticks. The bacteria can be spread by wind and dust, is extremely resistant to hot, dry conditions and can survive for a very long time in the environment. Pens can be disinfected with a chlorine bleach (0.05 percent hypochlorite final concentration), Lysol (1:100 solution) or a peroxide (5 percent) solution. (Source: from Q Fever fact sheets, the Center for Food Security and Public Health, 2007; and the Centers for Disease Control and Prevention, 2015.)

There are no symptoms of Q fever in goats, but it can cause late-term abortions, stillbirths or weak kids. The most common method of infection from goat to goat or from goat to human is through the placenta and birthing fluids. Bacteria can also be found in milk, feces, semen and urine. The spore-like form of the bacteria can survive for a long period after these fluids have dried up, which is why it is very important to dispose of placenta shortly after birth by either burying or composting. Note to raw milk drinkers: the Q fever bacteria can be found in raw milk.

To prevent the spread of disease, obtain advice from your local department of agriculture on the best way to dispose of placenta, fetuses and deadstock and related regulations. Fines can result for non-compliance.

Worldwide, the economic loss of production from dairy-goat herds to Q fever is substantial. An infection from Q fever in goats is heartbreaking on an individual basis as well. Spontaneous abortions can leave farmers with little to no milk production. When kids are born weak or stillborn, does can still produce milk, but it may be a limited supply. These does must then be re-bred and carry out full pregnancies before they produce milk again. The loss of kids also means a decrease in replacement animals and animals to be sold for meat in the upcoming season.

Always be conscious of the mode of transmission of this disease and work to keep routes of infection to a minimum. These include minimizing contact with animals from outside the goats' living space and disposing of livestock waste carefully.

Humans can contract Q fever, although most individuals (up to 60 percent) do not develop any symptoms. In approximately 38 percent of cases, Q fever causes acute flu-like symptoms two to four weeks after an individual is infected with the bacteria. Symptoms include high fever, headache, fatigue, muscle pain, sore throat, chills, chest pain and occasionally pneumonia. The illness usually lasts one to two weeks. Chronic Q fever is not common, but it is severe, with the infection lasting six months or longer. Usually it occurs in individuals who are immune-compromised or to animals that have pre-existing damage to their hearts. Medical attention is needed in acute cases.

Unfortunately, many doctors are not familiar with Q fever. If you need medical attention and suspect Q fever, make sure to let your doctor know you work with goats and have reason to think that Q fever may be the cause. Blood work can be done to conclusively test for this condition. (Source: Ontario Ministry of Agriculture, Food and Rural Affairs.)

Q fever checklist

Here is a checklist of precautions and steps to take to defend your family and your herd from Q fever. Following these procedures also helps you defend against many other illnesses as well.

- Wear protective gloves when working on an animal with an open wound, abscess or skin disorder, as well as during kidding. Disposable nitrile gloves can also be worn during milking.
- Try to make sure your goats kid while indoors in a pen that can be cleaned and disinfected both before and after giving birth.
- Wash your hands thoroughly several times a day, especially before

Injections

Learning how to give injections will save you a lot in vet bills. Ask someone experienced to give you a lesson and to watch over you to make sure you've learned how to do it correctly, using both methods, intramuscular (IM) and subcutaneously (SC). If you don't have such an acquaintance, add it to the list of questions at your annual vet visit. I prefer to make these injections in the neck, whether IM or SC, so as not to damage valuable meat with an improperly placed needle.

Intravenous (IV) needles are rarely necessary with goats. If a problem is serious enough to require an IV, leave the injection to the vet. You should also keep in mind that some injections, especially those from certain vaccines, can result in lesions on the goat if done incorrectly or in the wrong location. If you show or plan to show your goats, take extra care to avoid these imperfections. To avoid lesions from injections, ask your vet whether there is a vaccine that's less prone to creating a lesion; always use a new, clean, sharp needle; inject in an inconspicuous location; and restrain your goat so her movement doesn't cause more damage than necessary from the needle.

and after milking and before and after treating a wound, abscess or other condition on a goat. Wash your hands before eating, smoking, drinking, handling food or returning to the family home.

- Wash equipment to make sure it is free from animal manure, urine, milk and other body fluids when necessary and practical to do so.
- Footwear that has been worn to the barn should always be cleaned afterward, and clothing worn to the barn should be washed frequently with the hottest water possible. Adding chlorine bleach to the wash water when possible is recommended. Do not wear clothing or footwear to other locations, especially to another farm with livestock.
- Visitors to your farm should wear clean protective coveralls and footwear or foot coverings whenever possible. This applies especially to visitors from other farms.
- Animals such as pets and wildlife should not consume birth products or deadstock. Dispose of these according to your state or provincial regulations.
- Maintain a closed herd. This means do not purchase animals from other herds or lend animals out to other farms. Limit traffic of breeding bucks and quarantine new animals coming to your farm. Speak to your vet about quarantine procedures.
- Regularly clean out pens and disinfect when possible, especially after kidding.

- Consume only pasteurized milk and milk products.
- Pregnant women should avoid assisting or being present in the barn during kidding, especially if there have been unexplained abortions or stillbirths. Infants, young children, the elderly and those with weakened immune systems are also more susceptible to picking up diseases like Q fever from goats.
- Try to keep your barn and barnyard free from vermin such as mice, squirrels and rats. Wild cats or numerous barn cats can also transmit and carry disease. If possible, limit bird traffic into your barn, and do not house chickens and other poultry with the goats.
- Do not keep sheep and goats at the same time. These two species pass many diseases back and forth.
- Keep your barn and barnyard as free from flies as possible. Flies breed in damp, warm, wet locations. To control the fly population, keep pens clean, limit wet/damp areas and avoid feed and water bowl spills and leaking water lines.
- Keep a logbook of animal IDs, animal health history and kidding history. Record any thoughts, questions or concerns in this book.
- Ask your veterinarian to make an annual visit and herd inspection, and refer to your logbook during this visit. Ask any recorded questions and take down answers and advice from your vet. Call your vet for a special visit when a major event occurs such as an abortion, a stillbirth, mastitis or an unexpected death.
- Restrict the access of goats to manure storage facilities as well as access to buildings.
- Keeping air quality in mind, housing should be as dry and free from drafts as possible.

Metabolic Diseases

A host of other metabolic diseases—nutrition-related diseases—can affect goats. These can all be avoided through proper nutrition and feeding. Your feeding plan should be reviewed if there is a significant change in forage quality, such as a switch from first to second cut or from alfalfa to grass hay. The cost of paying for a proper nutrition plan will be less than the cost of treating a goat for a disease, or worse, losing a goat to disease.

Abortion
Abortion can be caused by several factors. The common diseases that

can precipitate an abortion in goats are Q fever, chlamydia, toxoplasmosis and listeriosis. Abortions may also be caused by nutritional deficiencies or by a goat delivering a severe blow to another goat's abdomen. Be sure to offer free-choice iodized salt to your goats at all times to avoid abortion from iodine deficiency.

Work with your veterinarian to determine if one of these diseases is the cause of abortion. To find a definitive explanation, fetuses and placenta have to be sent to the lab. If you have only one goat bred, you may decide it is not worth the expense of conclusively finding the cause. If a disease caused the abortion, the goat will fight off the disease before the next pregnancy. You may decide that it makes better sense to save your money on the vet bill (the mystery can't always be solved) and invest your efforts on re-evaluating your housing, bio-security and nutrition programs.

But if you have more than one bred goat, it may be well worth the expense in order to avoid additional abortions in your herd. Speak to your vet and make a decision based the estimate of cost and the probability of finding a conclusive result that suggests a possible remedy.

Acidosis

Acidosis is the term we use for a goat that has a decrease in rumen pH. Remember, the rumen is simply a vast fermentation tank that's full of bacteria. If the bacteria present in the rumen don't have the necessary conditions to keep them alive, they die off. Typically, the absence of ideal conditions for the bacteria is caused by a lack of adequate forages such as hay or an overabundance of concentrates such as grain or fresh spring grass.

These changes in diet cause changes in the delicate balance of species of bacteria present in the rumen. Acid-producing bacteria populations thrive. They also produce acid, which causes a decrease in pH that promotes further growth of acid-loving bacteria. In turn, rumen pH continues to drop, becoming more and more acidic. Eventually, the pH of the blood also drops, causing an overall pH drop in the body of the goat. The goat will die if the pH is not brought back to normal levels.

Again, knowing your goats well will help you quickly diagnose an acidotic goat. The key symptom is a failure to ruminate, or chew her cud. The goat looks listless and is most likely standing by herself. She may also be bloated, but she can be acidotic and show no signs of bloat. If your goat is bloated, she's most likely in pain and grinding her teeth. Grinding of the teeth may also be seen in a case with no bloat. Acidosis can be very painful as any humans with excess gas or acidic stomachs understand (although human causes are very different because we are monogastrics, not ruminants).

With this ailment, time truly is of utmost importance. Removing any concentrates (such as grain) and ensuring lots of good-quality, high-fiber hay helps if your goat is still up to eating. A drench of sodium bicarbonate (baking soda) may help. I use one-quarter cup of baking soda in a little more than two cups of water. The baking soda does not dissolve well (if at all), so shake the solution frequently. The goat may drink the solution herself, which is best. If she won't drink, use a 60 mL syringe to force the solution down your goat's throat. Tilt her head up, and aim the syringe at the back left side of the goat's mouth. Administer the solution slowly to avoid the goat aspirating on the solution. Make sure she swallows before releasing her head or she will most likely spit the solution out.

I have also given Pepto-Bismol (according to weight instructions on the bottle) with success. The goal is to increase the pH of the rumen back to a pH closer to neutral. If that can be achieved, the proper bacteria will once again populate the rumen. This takes time, and it is best to move your goat to a comfortable space where she can recover on her own. Fingers crossed that your goat will once again be able to consume hay on her own.

Acidosis is not a favorable condition, and death is often the result, even though prevention is relatively easy. Keep grains locked out of goat's reach, ensure plenty of good-quality forages such as hay, restrict access to fresh spring pasture, and when goats do go on pasture or before giving grains, be sure they have a full rumen of hay beforehand. Decrease stress in your goat's life as much as possible to keep the goat chewing her cud and the rumen functioning normally.

Medication

When treating a goat with any medication, record the goat's identity, as well as the date, dosage, circumstance and results. Do *not* consume the milk from this goat until a safe time has passed for the drug to disappear from the goat's system. That time depends on the medication used and the dosage given. Follow the manufacturer's directions or your veterinarian's advice. Many drugs used for goats are "off label," which means there are no dosages or withdrawal times specified for goats. Ask your vet for this information. This is one area where I wouldn't depend on the neighbor's word—or on mine, for that matter.

Bloat

This condition is preventable and treatable.

Bloat seems to occur when you least expect it. It can be caused by diseases such as enterotoxemia and ketosis or by nutritional factors such as a change in feed, a lack of water or overeating. A goat with bloat no longer ruminates or eats and shows signs of restless behavior, discomfort or extreme pain. She may lie down or stand and paw the ground. She may do a combination of the two by standing up and pawing, then lying down again. (This restless behavior is also consistent with kidding.) If you catch bloat early enough, you may be able to treat the goat with a baking soda and water solution, a commercial bottle of Bloat Aid or a similar product. I have had positive results with Pepto-Bismol as well.

With either product, I follow the directions on the bottle for weight and give orally with a syringe. The back left side of the mouth is the best place to administer medication orally. Raise your goat's nose up, pointing to the sky, to ensure she swallows. Be sure she does swallow the medication before releasing her head. Some vets say Pepto-Bismol doesn't work, but I have had great results with it in many different situations—especially with kid goats. Pepto-Bismol is easy to find, inexpensive and will not harm your goat even if not needed.

As with most goat ailments, it's best to prevent bloat rather than treat it. Preventive measures include making sure any feed changes are gradual, limiting your goats' access to fresh, lush pasture and always ensuring their rumens are full of hay before turning them on to pasture or feeding them concentrates such as grain.

If you're worried about bloat or acidosis because your goats are in a less than ideal state of feed or environment, offer free-choice sodium bicarbonate (baking soda) to your goats. Sodium bicarbonate is available at your feed store in large bags, and it is very inexpensive. Offering free-choice sodium bicarbonate also works to prevent acidosis. Always have free-choice fresh, clean water available to your goats, and be sure they have access to free-choice roughage such as hay, grass or straw. Remember that when you feed your goat, or any ruminant, you're also feeding the bacteria in its rumen. A change of feed can cause bacteria populations to change, which in turn causes changes to the rumen pH, gas and nutritional by-products. Controlled fermentation is critical, just as it is when you're making wine, beer or cheese.

Internal parasites

Internal parasites vary according to geographic location, housing conditions and number of goats. Consult your vet to develop the best, most cost-effective program for you and your goats. Today, there is a growing issue with parasite resistance to anti-parasite drugs, and your vet should have a good idea of what's happening in your area. At our farm, we used to treat our animals with a regular de-worming program. Now, because of resistance to these drugs, we treat them only on an as-needed basis.

When the Vet Visits

Think of your veterinarian as a partner in disease prevention rather than the person you're calling in an emergency. It's best to make your investment in veterinary care upfront and before a problem occurs. A veterinarian may see differences in your animals that you may not spot, simply because he or she does not see them every day.

Questions to ask your vet
- Can you evaluate and make suggestions about our facility and the level of our goats' comfort?
- Can you suggest a vaccination program?
- Can you suggest a de-worming program?
- Can you recommend a good nutrition plan for our goats?
- What mediations or treatments do you recommend for pink eye? For CLA? For lice?

There may be other questions you would like to add to the list. The important point is that you should do your homework ahead of time and anticipate any potential problems so you know what to ask the vet.

When calling a vet out for a specific incident, record all the information that your vet shares. If you've taken good notes and the same problem occurs later, chances are you won't have to call the vet a second time.

There are many diseases and illness that can affect goats, but an ounce of prevention is worth a pound of cure. Keeping goats is all about management—careful observation, experience and record keeping. All are necessary to become a knowledgeable and successful goat keeper. Enjoy it.

Keeping livestock is a choice. There will be ups and downs; learn from the downs and celebrate the ups. And strive for more ups than downs.

CHAPTER 8

Cheese Making

MAKING CHEESE WAS never an ambition or dream for me. I stumbled into it out of sheer necessity. Once we'd brought home our first goats in the spring of 2001, I had surplus milk from two high-producing does, and I simply couldn't bear to see it go to waste. What else can you do with eight liters of milk a day but make cheese?

Today, I am hooked on cheese making. Over the years, I've learned that cheese making is far more complex than the straightforward task of following a recipe. It is a process where perfection is always in sight. Even in my best cheeses, there is always something that can be improved on. That's the challenge. Then there is the reward of actually eating the cheese and tasting every perfect, but flawed, morsel. There is always a bit of each.

This chapter only includes recipes for fresh cheeses. Trust me when I say that mastering these fresh cheese recipes will set you up to be successful at making cheeses that are to be aged.

By making the cheese recipes in this book in the order in which they appear, you'll start to understand your milk and the process it goes through on the journey to becoming cheese. When you make an aged cheese, the end product is only as successful as the quality of the green cheese, which has yet to start the aging process. Green cheese is waiting for microbes to take over and transform the cheese to its final appearance, texture, aroma and character. All cheeses at the green stage are fairly similar in taste and appearance. Yet it is the quality of the green cheese that will largely determine the end product of the cheese after aging. Even the most perfect aging caves cannot turn out good cheeses if good cheese doesn't enter the cave. So let's start by making well-executed fresh cheeses.

Setting Up for Success

Checking and washing equipment

I know you know how to wash things, but washing goat milk off of stainless steel is a totally separate skill. Washing goat milk from plastic is even more difficult. (That's why I recommend stainless steel.) If equipment is not washed properly, the population of unwanted bacteria in your milk will increase.

It's entirely possible for the equipment to look clean when in reality it's absolutely filthy. When stainless steel is clean, truly clean, it gleams. You should feel the need to squint when you look at it. Even though I don't recommend using plastic, if you're using it, it shouldn't feel at all slippery. The surface of the plastic shouldn't be dull or hazy looking. Plastics are highly prone to scratches. To a microbe, a small, shallow scratch in plastic is like a valley in the Rocky Mountains. Thousands of microbes can hide in there. Scratches, pits or hollows in stainless steel are the same.

When examining the equipment, it's best to do so first with the naked eye. Note any areas of dullness or haze. Protein build-up appears as a rainbow haze on stainless steel and is extremely difficult to see on plastic. Fats on stainless steel appear where water beads up. If water doesn't evenly shed off the equipment, you have a fat build up. Plastic feels slippery or greasy when a fat build-up is left behind.

After you have examined the equipment with your naked eye, shine light on the equipment with a bright flashlight. Hopefully, your light gleams back at you from your gleaming stainless steel pot. If your pot has not yet had milk in it, there's a good chance your stainless steel is gleaming. If you have had milk in the pot before, I am going to guess that you'll see some kind of orange to brown haze.

This check should be done regularly—at least once a week. Always record the date and your observations in your journal. It's easy to become lax on the inspections, especially if you're making cheese and washing the equipment every day or even two or three times a day. Don't be fooled into thinking the equipment is clean. A visual inspection with the naked eye should be done every time you use the equipment before you sanitize it.

For cleaning, I recommend going to a company that supplies dairy farms with cleaning chemicals for pipelines and milking equipment. These chemicals are specially formulated for removing milk fats and proteins. Don't let the term "chemical" scare you. I know people who use common dish soap or commercial hand soap to clean their milk equipment. I also know that these individuals make poor quality, inconsistent cheese and can't figure out why. It's quite obvious to a knowledgeable outsider that the problem is poor hygiene. If you want good cheese, you must clean the equipment properly, following the correct procedure, and inspect the equipment regularly. If you don't, you risk losing an entire batch of cheese.

What you need to wash properly
- A vast supply of hot water.
- A large sink or basin in which to wash the equipment.
- Rubber gloves to protect your hands from the basic and acidic cleaners. Thick rubber gloves are a bonus because they also protect your hands from the heat of the hot water.
- Safety glasses to protect your eyes from splashing acidic or basic wash solutions.
- A detergent that will remove fat and protein. (Talk to an experienced chemical supplier to dairies and explain that you're manually washing cheese-making equipment.)
- An acid solution rinse that will neutralize the basic properties of the detergent. (The acid will remove any hard-water stone or milk stone that may be present on the equipment.)
- A variety of plastic scrubbing brushes. (Flexible but strong bristles work best. For small, out-of-reach areas, you may wish to invest in a test tube or baby bottle cleaner.)
- Some micro-fiber, anti-microbial cloths. These are special cloths that, because of their construction, do not promote bacterial growth. They generally have an open weave and dry very quickly. Remember, bacteria love moisture and need a feed source. Do not use sponges.
- Lots and lots of elbow grease.

Cleaning steps

1. Rinse and scrub off any visible milk or cheese residue with warm water—do not use hot water. Hot water will bake milk proteins onto the surface of the equipment. If curd is "stuck" on the equipment, scrub it off. Again, do not use hot water. The equipment should look clean before you move on to step two.
2. Fill your sink or basin with hot water—the hotter the better. Add the detergent to the water according to the manufacturer's instructions.
3. Wear your gloves and safety glasses. Place as much of the required equipment as possible into the sink. Make sure you still have ample room to wash. As you wash one piece, you can add another. Soaking will reduce your scrub time.
4. Scrub each piece of equipment thoroughly and many times over. When you think a piece of equipment is clean, it's not. Put it back in the sink to soak while you scrub another piece, and then scrub it again.
5. When you think that the equipment can handle no more scrubbing, you can rinse with hot water. Take care that your hot water with detergent doesn't become diluted with rinse water. If possible, rinse into a separate basin.
6. After all equipment and work surfaces are washed and rinsed, let the wash water out of the sink.
7. Rinse your sink well with water and refill with warm water. Add the acid rinse according to the manufacturer's instructions. Place all the well-washed, well-rinsed equipment into the acid solution. Give everything a quick scrub and rinse with tepid water. The equipment is now clean.
8. Place all the equipment in a well-ventilated area so that it can dry. Do not place on any material that can hold moisture such as a dish towel or wood. Do not nest equipment. Store equipment in a dust-free environment. Once equipment is completely dry, it can be stored in a large covered plastic tub.

Sanitizing equipment

All equipment must be sanitized before use. I like using chlorine to sanitize. It is effective and can be used at a low enough concentration to not burn the skin. Chlorine also becomes inactive as soon as it comes into contact with organic compounds such as milk. The one caution here is that chlorine will inactivate coagulant, so rinse any vials or containers well that will come into direct contact with your coagulant.

I am adamant about having a sink, basin, bucket or other container filled with a sanitizing solution at all times when making cheese. If you have to touch something else that may not be "cheese clean," you can dip your hands into the solution to be sure you won't be contaminating your cheese. If you drop a spoon on the floor, allow it to soak in your solution.

Follow the instructions on the bottle of sanitizer, fill a basin or large sink with a solution and don't be afraid to use it at every opportunity. At the start of your make, when you're preparing and gathering the equipment, toss it all into the sanitizer. Grab the equipment from there. Any equipment, such as your thermometer or scale, that cannot be submerged into a liquid, wipe with sanitizer before using.

These washing and sanitizing instructions can be used for milking equipment, pasteurizing equipment and cheese-making equipment. By following them, you'll automatically be a successful cheese maker. Good cheese requires high-quality milk and clean equipment. Clean, clean, clean—from hands, to udders, to pails, to spoons.

Calculating Quantities

My recipes call for 10 liters (or quarts) of milk and six teaspoons of lemon juice. Let's say you have only four liters of milk. So how much lemon juice do you need? To calculate the amount, enter your equation in your journal. Make sure the numerator (above the line) and denominator (below the line) are in the same measurement: quarts and quarts, and teaspoons and teaspoons.

$$\frac{10 \text{ liters milk}}{4 \text{ liters milk}} = \frac{6 \text{ teaspoons lemon juice}}{x \text{ teaspoons lemon juice}}$$

"x" is the unknown amount for which you're calculating.

To solve the equation and determine the amount of lemon juice you need for four liters of milk, multiply the two diagonal numbers.

In this case,

four liters × six teaspoons = 24.

Since you're trying to find the value of "x," you then divide 24 by 10, which equals about 2.5.

Therefore, you will need to use 2.5 teaspoons of lemon juice if you're using four liters of milk.

Everything that comes into contact with milk should be properly washed and sanitized before use. Otherwise, you'll be setting yourself up for failure.

Any good cheese maker will tell you that cheese making is 80 percent washing and 20 percent making cheese.

The author stirring a pot of feta curd.

Cheese Curd

There are three ways of getting milk to form curd. Curd can be produced by heat precipitation, acid coagulation and enzymatic coagulation. Heat-precipitated cheeses are generally a rubbery, firm curd without complex flavors. These cheeses are still delicious but come with simple personalities—think of ricotta, or the Indian-style paneer. Curd that's formed by acidification is very soft and typically creamy, with a shelf life that's measured in days or weeks. Think of a creamy chèvre. Curds formed by enzymatic action or rennet are harder cheeses that can be aged for weeks, months or even years. These cheeses have complex flavor profiles that often linger on your palate and in your thoughts. You can pick out more than one or two flavor characteristics in these cheeses. Think cheddars, Gouda, Havarti, or even feta.

Recipes

Ricotta

Ingredients
- **10 liters/quarts raw goat milk.**
Two- to three-day-old milk is perfect for this cheese because the bacterial and enzymatic action in the standing milk has already started to lower the pH of your milk (pH is a numeric scale used to specify the acidity or alkalinity of an aqueous solution). You may use raw milk because the milk will be taken to extremely high temperatures. Microbiogically speaking, this is a very safe cheese to make.

    For this cheese, you can also use whey or a whey/milk combination. Your cheese yield will be higher with a higher concentration of milk. I'm including this recipe is to explain how milk behaves in different situations. For the beginner cheese maker, start with 100 percent milk and, as you gain experience and create whey in your other cheese makes, you can attempt this cheese using the whey.
- **30–60 mL (2–4 tablespoons) lemon juice or white vinegar.**
The addition of an acid such as lemon juice or vinegar may not be necessary. Depending on your milk, curd may separate out with just heat. If you don't see curd forming, adding an acid will speed up the process. A fun experiment: try making your cheese by using just heat. Once all the curd is removed, add lemon juice or vinegar. See how much more curd you get.
- **Salt.**
Ideally, cheese-making salt is always used to make cheese. This is pure salt; no iodine, minerals or other contaminants. In some areas, this type of salt can be difficult to find.

    The next best variety is kosher or other coarse salt. This salt tends to be quite pure. The disadvantage (or advantage) of this type of salt is that it has a coarse texture that doesn't incorporate nicely into a fresh, creamy cheese such as chèvre, but does do well for dry salting a harder cheese. Kosher salt can be put through a salt mill or a pepper grinder. Flaked salt, if you can find it, is ideal for dry salting a firmer cheese. When dry salting, fine salt can wash away easily with the release of whey.

    Table salt contains iodine, which will inhibit microbial growth. That's why you can use it in a fresh cheese such as chèvre, where you don't want microbial growth.

    Neither rock salt nor sea salt should be used in cheese making. These types of salt come with impurities. Minerals that affect pH, microbes such as molds and yeasts, and bits of sand can all be found in them. Feel free to experiment but, as a general rule, these trendy salts have no place in cheese making.

Materials
- Journal and pen
- Double boiler (see description in Pasteurization section on pages 100–103)
- Long-handled stainless steel spoon
- Long-handled stainless steel slotted spoon
- Teaspoons with metric measurments marked on spoon
- Thermometer if desired (a dairy thermometer that clips to the pot is nice but not necessary)
- Stainless steel colander and bucket or large bowl to catch the whey (plastic or glass can be used for your catching vessel, but stainless steel is ideal)
- Cheesecloth
- Large glass or stainless steel bowl

Method
1. Record date, the type of cheese and any relevant details about the milk (age, the goat the milk is from, diet of goat, anything of interest to you).
2. Wash hands and sanitize the equipment and work surfaces.
3. Fill the larger pot of the double boiler with water. (Tap water is fine for the larger—bottom—pot, but it may contain chorine, which is unstable. So be sure tap water doesn't come into contact with your milk or cheese.) Place double boiler on stove with burner on high. Wait until water is almost to the point of boiling.
4. Fill the smaller pot of the double boiler with milk, and set it inside the water pot. Leave burner on high heat. The milk is protected by the water jacket and will not burn, though it is heated to a very hot temperature, 90°C (194°F) or more—almost to boiling. Stir milk constantly to create a consistent temperature throughout the milk, but do not let it boil.

Cheesecloth

Traditional cheesecloth is a white, airy and lightweight muslin fabric, but that cloth isn't effective for the purpose of draining curd from whey. The best fabric is cotton with a loose weave. Bed sheeting or pillowcases work well, whether new or old. Wash thoroughly and sanitize before using. My favorite bags are made from inexpensive cotton dish towels. My mother sews two of them together to make a bag, and we use these in our home as well as in our commercial cheese plant.

5. As the milk reaches a temperature close to 87°C or so (189°F), reduce heat to low. If you're not using a thermometer to monitor temperatures, note when the milk starts to form small bubbles on the side of the pot. It will start to steam shortly afterward. Just as the milk starts to steam, reduce heat to low.
6. If desired, add lemon juice or vinegar, one tablespoon at a time. You should start to see some white clumps floating around in your milk. Alternatively, wait to see if curd forms from being heated, then add the lemon juice or vinegar if curd isn't forming or if you'd like more curd from your milk.
7. Keep milk at a high temperature, stirring constantly. Add additional lemon juice or vinegar until the color and texture of the milk changes from a white milky liquid to a greenish-yellow watery liquid with lots of white curd.
8. At this point, turn off burner and remove milk pot from double boiler.
9. Line colander with cheesecloth, and place colander over the collection bowl or bucket.
10. Pouring the hot whey and curd directly into the lined colander can be dangerous because hot whey splashes. Instead, first spoon curd carefully into the colander. Then pour the lump-free whey, which will still have small bits of curd in it, over the curd. (Cheese yields are never very large, so you should recover all the curd you can.)
11. In the colander, gently toss curds to remove as much whey as possible. Taste the curd to determine if you need to add salt. Make tasting notes in your journal. As you become more experienced, you'll be able to taste the differences in your curd. Such differences may be a result of milk seasonality, milk-to-whey ratio or the amount of lemon juice or vinegar. For this type of cheese, microbes won't be a factor until the cheese ages. Sprinkle salt over the curds.

When the cheese has cooled to room temperature, place the curds in a container with a tight-sealing lid. Refrigerate. Ricotta should last three to five days in the refrigerator. As always, remember that fresher is always better. This cheese also freezes nicely, so you can have a supply of cheese on hand during the months when your goat is not producing milk.

Salt

Salt has long been used as both a preservative and a taste enhancer. When added at the right time and in the right amounts, it can also influence the moisture content of your final product. By controlling moisture and microbe growth, you're also controlling the cheese's pH. Once you've tasted unsalted curd, you'll understand the value of salt.

Remember, too, that when it comes to salt, your sense of taste can change from day to day. Levels of hydration, foods eaten recently and general health (a common cold affects our sense of taste) can cause changes in the way your body interprets taste. Ideally, always ask a few family members or friends to provide second opinions.

Making cheese doesn't demand an investment in costly equipment such as pH meters and scales. Nor does salting cheese have to be an exact science. However, as you become more adept at making cheese and attempt cheeses that grow molds or bacteria on them (Camembert, blue cheeses or washed rinds), salting does become more critical. For the simple cheese recipes included in this book, salting cheese to taste is sufficient.

Creamy Goat Cheese

Our goal when making this cheese is to decrease the pH of the milk to 4.6 or less, at which point a curd will form. (A solution with a pH of less than 7 is acidic; a solution with a pH greater than 7 is alkaline, or basic.) The chemistry that explains how a lower pH causes this change in the milk protein is too intricate to describe here, but I'll offer a rudimentary explanation. The protein molecules in milk stay free-floating because the protein carries a negative charge. Since like charges repel one another, the protein molecules do not clump together. However, when an acidic solution, such as lemon juice or vinegar, is added to milk, it causes the protein molecules in the milk to lose their negative charge. Once the protein molecules no longer carry a negative charge, they cannot repel one another, and they clump together and form a curd.

The acidification in the following recipe is achieved with the addition of lemon juice or vinegar. You can play with quantities of both substances, and each one will produce a different flavor in your cheese. Vinegar tends to be slightly more acidic than lemon juice, and you may find that vinegar makes curd more quickly than does lemon juice. The curd in this recipe is formed by the acidity of the milk, rather than heat precipitation as in the ricotta recipe.

Ingredients
- **10 quarts/liters pasteurized* or raw whole milk.** Milk will be pasteurized during this cheese make, so starting with raw milk is acceptable.
- **10 large lemons, squeezed and strained of seeds, or 1 cup of vinegar**
- **Salt to taste**

Materials
- Double boiler
- Large glass or stainless steel bowl
- Long-handled stainless steel spoon
- Stainless steel colander that fits on top of bowl or bucket to collect whey
- Cheesecloth and string
- Thermometer

*In addition to pasteurizing milk at 62°C (144°F) for 30 minutes, pasteurization is achieved at 72°C (162°F) for 16 seconds. This method doesn't work well for fluid milk because it yields a cooked, flat taste. However, in this cheese make the milk becomes naturally pasteurized during the process because the heat required exceeds a pasteurization temperature and time of 72°C (162°F) for 16 seconds. This means that your finished product will be safe to consume. High moisture levels and a low salt content make fresh cheeses like this one potentially dangerous to eat but, with the method I use, the cheese becomes naturally pasteurized during the cheese-making process.

Method
1. Record date, type of cheese and any relevant details about the milk, such as its age, the goat(s) who supplied it, feed, or any other factor you feel may change the quality or taste of the final product.
2. Fill larger chamber of double boiler with water, and place over high heat.
3. Add milk to the smaller pot and place inside water jacket.
4. Stir milk, monitoring temperature, and bring milk up to 74°C to 76°C (165°F to 169°F). At this stage, the milk is pasteurized.
5. Turn off heat and add lemon juice or vinegar one-quarter cup at a time. Stir for 10 seconds.
6. Let milk settle for 5 to 10 minutes or until it separates into stringy white curd and greenish-yellow liquid whey. If nothing has happened after 10 minutes, add another one-quarter cup of lemon juice or vinegar to increase the acidity of the milk. If after several additions your measured quantity of lemon juice or vinegar has been used up, turn the heat back on and gently heat the milk till you see curd and whey separation. Once it begins, turn off the heat and let milk sit for 20 to 30 minutes.

Fresh, creamy cheeses can be dressed up in many ways with many different ingredients.

7. Line colander with cheesecloth, and place colander over a bowl or bucket.
8. Pour curds and whey from the milk pot into colander.
9. Let curds drain for several minutes until most of the whey has passed through into the container below.
10. Fold up the corners of your cheesecloth, and tie with string.
11. With the use of a stick or similar implement inserted through the knotted bag, hang the cheese bag over a bowl or pail. You may have to be a little inventive with this step. If you have pets, make sure they can't get at it. Cheese should hang for an hour or so.
12. Remove cheese from the cloth when it's at the desired consistency. You should end up with a spreadable cheese (however spreadable you like it). The cheese will become firmer when refrigerated. Keep that in mind when deciding when to stop draining the cheese.
13. Salt to taste—1 percent salt by weight is a good place to start. You can adjust amount according to taste. Add herbs or spices if desired.

Keep cheese in a tightly sealed container in the refrigerator for up to one week. This cheese also freezes nicely.

Chèvre

There are two ways to acidify milk. One occurs very quickly with the addition of lemon juice or vinegar, as in the recipe for Creamy Goat Cheese. The other is to let the bacteria that are present in milk do the work. As these bacteria consume lactose, it becomes lactic acid. This lactic acid performs the same function as lemon juice or vinegar. The difference between the two is that it takes time for the bacteria to grow, multiply and convert enough of the lactose so the milk reaches a pH of 4.6. At this pH, a soft curd is formed. But if the pH is too high, at 4.7 or above, it is difficult to drain the whey from the curd. If the pH is too low, 4.5 or less, the cheese has a grainy, crumbly texture. With chèvre, we want a creamy cheese that isn't too wet because it contains extra whey.

When making chèvre, you can control the acidification of your milk and resulting curd by controlling the amount of bacteria you add to the milk, the temperature at which you keep the milk and the time during which the bacteria are held at this temperature. With this recipe, you'll need to master the growth rate of the bacteria. As always, this will require keeping careful notes about your milk, curd and resulting cheese.

Before we get to the how-to of making chèvre, you'll need to acquire

Draining Cheese

Practice makes perfect when draining cheese, so record your results in your journal—the hanging time, as well as the room's temperature. Temperature determines how fast the cheese drains—the warmer the room, the faster. In a cold room, or if you choose to refrigerate the cheese, the draining process will be slower. Keep that in mind if you need to go away after hanging your cheese, or if your cheese needs to drain at night and you don't want to wake up at 3 in the morning to deal with it.

the bacteria that make the magic happen. In the cheese-making world, we call this *starter culture*. Starter culture is used in acid-coagulated cheeses such as chèvre, as well as enzyme-coagulated cheeses such as cheddar. You have to purchase starter culture from a cheese-making supply company. (While you're at it, you should purchase some rennet and calcium chloride at the same time.) Be sure to ask for a culture that works well with acid-coagulated cheeses such as chèvre.

Starter culture is measured in units, and these should appear on the packets you purchase. Every culture has a different amount of active bacteria per gram, and it's important to know the number of bacteria added in order to create cheeses that are consistent from make to make. Danisco (the manufacturer of the cultures we most often use) measures its cultures in DCUs. (DCU, which stands for direct culture unit, is a measure of the activity of the powder in the packet.)

The amount of DCU (or other units) that you get from a culture company is only a guideline. That number is based on a standardized milk which has a specified amount of fat or protein.

Ingredients
- **10 liters/quarts pasteurized whole milk** (see instructions for home pasteurization on pages 100–103). I find chèvre is easiest to make and tastes best with end-of-summer milk.

Making Cheese Safely

Milk at room temperature is the perfect environment for bacteria to thrive in, and this includes the pathogenic bacteria that can make you very sick. For this reason, all soft, fresh cheeses with a high-moisture content, close-to-neutral pH and low-salt content should always be pasteurized. It's possible to become extremely ill from soft, fresh raw milk cheeses. In my view, it is simply not worth the risk. Educate yourself, know how to produce good-quality milk and how to wash and sanitize equipment, and learn about the cheeses that can be made more safely with raw milk.

- **Starter culture.** Well before you start, determine how much culture you need to use and how you will measure the culture. If no guidelines were given regarding the amount of culture to use with a given volume of milk, contact the supplier who sold you the product and ask for assistance.
- **Coagulant (optional).** The instructions on the package of coagulant are intended for harder enzyme-coagulated cheeses. Only a drop of coagulant, if any, is needed with soft, acid-coagulated cheeses. While coagulant helps with drainage, too much can have the opposite effect. You may decide not to add any coagulant at all until you're more familiar with this recipe.
- **Calcium chloride (optional).** Calcium chloride is added to strengthen the milk's ability to set. Follow the instructions on the bottle, although the rate will depend on the concentration. If there are no instructions, call your supplier. Soft, fresh acid-coagulated cheeses need much less than what's called for with harder enzyme-coagulated cheeses.
- **Salt to taste** (1 percent by weight is a good place to start).

Tips—Making Chèvre

- The warmer the temperature, the faster bacteria multiply. When working with a temperature range of 26°C to 30°C (79°F to 86°F), for instance, bacteria will multiply faster at the warmer end of the range.
- The higher the fat content of your milk, the more starter culture you need to add. You'll achieve better results if you slightly increase the amount of culture when making cheese with fall or winter milk.
- Rennet, or coagulant, can be added in small amounts to assist with drainage; however, pH plays a bigger role than rennet in drainage speed. The lower the pH, the faster the whey will drain and the more crumbly or grainy your chèvre will be.
- Growth rate of bacteria dictates the acidification rate. Learn to control the finished cheese by controlling the bacteria growth in the vat. Keep careful notes and pay attention to fine details such as room temperature, vat temperature, bacteria amounts and type of culture used.
- Measure starter culture accurately. The smaller the amount of milk, the more critical and difficult it is to measure the starter culture is. Use a scale that measures in small amounts.

Materials
- Journal and pen or pencil
- Stainless steel double boiler
- Long-handled stainless steel spoon
- Teaspoons with metric measurements indicated
- Thermometer
- 1–2 cheesecloth bags (one for backup in case of breakage)
- Bucket or sink filled with sanitizing solution
- A stainless steel scoop or measuring cup, with a pouring lip (plastic will do)
- A warm, draft-free environment in which to place the cheese pot for 18–24 hours
- A basin or bucket in which to drain the chèvre bags

Method
1. Record date, the type of cheese and any relevant details about the milk, such as its age.
2. Wash hands, and sanitize all equipment and work surfaces.
3. To save time and energy, make the chèvre right after you have pasteurized your milk. It will mean less washing of equipment, and you won't have to cool your milk only to heat it up again. After pasteurization, cool your milk to a standing temperature of 27°C to 30°C (80°F to 86°F). You don't want the milk to fall below 27°C (80°F), so if your room is on the cooler side, aim for a higher temperature. It's better to be too warm

than too cold. Keep the water in your double boiler at 27°C to 30°C (80°F to 86°F) as well. Its insulating and heat-holding power will help milk remain at temperature for a longer period of time.

4. Once milk has stabilized to the desired temperature, add starter culture, calcium chloride and coagulant (if using). Stir for about two minutes, longer for larger volumes of milk. Remember to record the temperature, time, amount and type of culture and whether you added coagulant and/or calcium chloride. You'll know your vat has reached a stable temperature if, after two minutes of stirring, the temperature of your milk has not changed.

5. Move double boiler and milk to a warm and draft-free location for the next 18–24 hours. Choose your location well as you don't want to move the pot during this phase.

6. Cover double boiler with a lid. If the room temperature is cool or drafty, wrap a towel or blanket around your double boiler to help hold heat. If you don't have a warm place to let your milk rest, you can always place your double boiler in a cooler. (But don't let the milk get too warm. The bacteria in the milk will create some heat throughout the process.)

7. Don't add heat to the chèvre during incubation. Go with what you have and record your observations. Make changes for your next batch if necessary. If you're working with previously pasteurized cold milk, heat the milk to the desired temperature. Record the temperature, the time and whether you used warm milk right after it had been pasteurized or cold, re-warmed milk.

8. Let milk rest, undisturbed. At the 18-hour mark, check on the curd and continue to check every hour after. The milk should change from a liquid to a solid, gel-like mass with a layer of greenish-yellow whey on top. As a test, I rub the top of the surface of the milk/curd with the back of my finger. If the curd is set, my finger glides over the surface of the curd. If the milk is almost set, my finger dips into the milk and emerges coated with thick, white milk. If your milk has made a nice firm curd but you don't see any water like whey on the surface, leave the curd for another hour. It won't hurt the curd, and the next time you look, there should be whey present.

9. Time to hang! Place cheesecloth bags in sanitizing solution for at least seven minutes. During this time, scoop off as much whey as

Cheese Making

White cheese curd set in a pot with surrounding yellowish whey.

Preparing chèvre for hanging by scooping into a cheesecloth bag.

you can from the top of the cheese curd. Gently press down on the curd to fill your scooping container with whey. Pour the whey into a clean, sanitized bucket or other container, and reserve.

10. Rinse sanitized cheesecloth bags in the reserved whey. The whey deactivates the chlorine and helps the curd drain a little better. Don't worry if there doesn't seem to be enough to rinse the bags—it doesn't take much.

11. You now have two options. One, stir the curd and whey with a spoon and then scoop it into cheesecloth bags. Or two, scoop or ladle curd into the bags without stirring. Depending on the pH of your curd, stirring makes the curd drain either slower or faster. With a relatively lower pH, the curd drains more quickly without stirring. If the curd has a relatively higher pH, the stirring really slows down the curd's draining time. For this reason, it's probably best not to stir the curd until you're familiar with the cheese you're making (you can never go wrong by not stirring).

12. Once you're more comfortable with the recipe, feel free to experiment by stirring the curd. Just remember to record the results in your journal. Stirring can change the consistency of your cheese and slow the drain time.

13. Scoop all the curd into cheesecloth bags.

14. Hang bags in a warm location to drain. Collect and reserve whey.

15. Chèvre can take from five to 24 hours to drain, though ideally

Whey dripping from chèvre curd. Bag is suspended by a kitchen cabinet knob over a turned-off stove.

Fresh chèvre curd after being turned out of the cheesecloth bag. Time to add salt!

draining time is eight to 12 hours. Each person's taste for creaminess is different. Hang the cheese for less time if you want a very creamy, spreadable cheese. If you want cheese to roll into logs or another shape, hang it for longer. If curd drains too quickly, the pH of the curd is too low. If curd drains too slowly, the pH of your milk is too high.

16. You may also wish to place your strained curd into cheese molds or even put whey and curd into cheese molds instead of draining your curd in bags. It's up to you. Have fun and remember to record everything so you can use the information for future cheese makes.

17. Once you're happy with the consistency, it's time to add salt and other flavorings. If you're using a scale, start with 1 percent salt by weight. As a guideline, a 10 liter/quart batch of milk yields roughly two kilograms (4.5 pounds) of cheese. In this example, you need a total of 20 grams (one tablespoon) of salt to salt the whole batch. If you're not using a scale, start with slightly less than one tablespoon and increase salt if necessary. Remember, you can always add more salt, but you can never take it away.

18. After cheese is salted and mixed, the real fun begins. Add herbs or other flavorings such as lemon, figs or cranberries. Use your imagination here. Sweet or savory, or you may wish to leave your cheese plain. Basil, thyme, dill, garlic, garlic scapes, lavender, lemon, curry, cinnamon, maple syrup,

Chèvre flavoured with cranberries and lemon zest.

fruit liqueurs. Or try dried fruits such as apricots, figs, cranberries or dates. Fig and lemon is a delicious combination. The possibilities are endless. Shape your cheese into balls, logs, upright cylinders or small disks. You may choose to roll your creations in herbs and other flavors instead of incorporating them into the cheese. You may wish to store your cheese in small covered containers and use the cheese as a quick snack.

19. If chèvre is refrigerated, its shelf life is typically at least one month, but longevity ultimately depends on how much care you take in the process. If yeasts, bacteria and/or molds have been allowed

If Your Curd Doesn't Set

If your curd has not set after 24 hours, don't add coagulant in an effort to make a different cheese—the milk is far too acidic for coagulant to work. But don't throw the cheese away. Enjoy the fact that you now have a supply of beautiful cultured milk which you can use in smoothies; in baking recipes that call for sour milk, buttermilk or sour cream; in soups or as a low-fat substitute for sour cream in gravies and on baked potatoes. You can even add maple syrup, honey or brown sugar to it and use as a dessert topping on fruit, fruit pies and coffee cakes. Be creative. This cultured milk also freezes well. If there's some solidity to your curd, you can try hanging it to see what you get. You may be pleasantly surprised.

Increasing Batch Size

To increase your batch size and use up some older milk, you may be tempted to add cold pasteurized milk to warm just-pasteurized milk. However, adding cold milk to warm (or vice versa) will cause the milk to release some of its fat, which, depending on the milk's fat content, you will most likely see as oil droplets on the surface of the milk or curd. If possible, warm the cold milk first before adding to the warmer milk. You can add cold milk to warm, but note that you may see a slight drop in cheese yield because of the release of fat into the milk. The addition milk may also change the final texture of your cheese.

to contaminate the cheese during the make, it won't last as long. Watch for small colonies of microorganisms growing on the surface of your cheese. Molds are fuzzy and are usually shades of gray, blue, white and occasionally black (Note: the short black mold, similar to what grows in damp window sills, is a dangerous mold you need to avoid.) Yeasts appear as shiny smooth bumps on the surface of your cheese. Visible bacteria on your cheese tend to be shades of orange, red or brown (*Brevibacterium linens*) or fluorescent pink or yellow (*Pseudomonas sp.*).

Paneer

Also called curd, fresh cheese, or farmer's cheese, paneer is an important and versatile ingredient in a lot of Indian lacto-vegetarian recipes. This cheese is not aged and does not melt when heated. I love paneer cut into crouton-sized cubes, fried in oil and garlic until browned and served as a replacement for bread croutons on salad. Fried paneer cubes are also delicious in rice, peas or other vegetables. Think of it as a substitute for tofu.

Ingredients
- **10 liters/quarts milk**
- **Juice from 10 large lemons, strained of seeds**
- **Salt to taste**

Materials
- Stainless steel milk pot from the double boiler
- Colander lined with cheesecloth over bucket or bowl underneath to catch whey
- Stainless steel spoon
- Plate that fits down inside the colander. This will be used to press the cheese curd
- Small jug of water (or another object) to add weight to your cheese during pressing

Method

1. Record date, the type of cheese you're making and any relevant details about the milk, such as its age.
2. Put the milk in the stainless steel pot and place over medium high heat. Stir continuously until the milk boils. Take care that the milk does not burn.
3. Once the milk is thoroughly boiling, remove from heat and add lemon juice. The milk will start curdling right away. You will see curd and greenish-yellow whey.
4. Carefully scoop and pour the curds and whey into the cheesecloth-lined colander.
5. The curds will remain in the cheesecloth. Allow to sit until all the visible whey has drained from the curd. Gently tossing the curd will help in the curd draining.
6. Sprinkle a small amount of salt over the curd. Paneer can be made without salt, but a pinch added helps with the taste and the paneer should not taste salty.
7. Fold the cheesecloth around the cheese. Press the curd with your hands or the back of a spoon. The goal is to squeeze every last bit of whey out of the curd. Place a plate on top of the curd (which is still in the colander). Place the jug of water (or other weight) on top of the plate. Leave for several hours or overnight.
8. Unwrap your cheese. It should be a solid block. It may not be an ideal shape, but that's fine—and part of the fun of cheese making. Now that you know how to make the cheese, you can experiment with different shaped containers to press your cheese. The only requirement is that the container needs holes in it to drain properly.

What to Do with All the Whey

Don't throw it out. There's a lot of goodness in whey. It can be fed to animals such as pigs, dogs, cats and even goats. We feed our whey back to the kids we are raising for meat. I don't like feeding the whey to the female goats because we want to milk them and I don't want the does to get a taste for milk. I have had does drink milk from their own udders. This isn't very productive if you want the milk for yourself.

Use the whey in the kitchen. Add extra nutrition to rice and pasta by cooking them in whey instead of water; you may end up with ricotta for your pasta or rice as well. Steam your vegetables in whey. Use it for broth for soup, or in place of buttermilk or sour milk in muffin recipes. Again, use your imagination.

Feta

The curd of this cheese is made by adding an enzyme that coagulates milk—a process different from heat-precipitated or acid-precipitated curd. This cheese is a fun one because you actually get to cut curd. It also requires the purchase of rennet. Rennet can be hard to find. I have had luck getting it (or some other form of coagulating enzyme) at drug stores (talk to the pharmacist) or health food stores. If you have a farm nearby that makes cheese, they may sell some coagulant to you.

In addition to the rennet used to form the curd, a starter culture will be used in this recipe to ripen the milk before the rennet is added. The bacteria in the milk will mainly consume lactose and convert it to lactic acid. This decreases the pH of the milk, making it slightly more acidic.

Ingredients
- **10 liters/quarts whole milk.**
- **500 mL (2 cups) live culture yogurt** (one tablespoon per liter/quart of milk). You can purchase a powdered yogurt or feta culture from a cheese-making company and follow the manufacturer's instructions, or you can use any plain live culture yogurt. You'll find a yogurt suitable for yogurt making at your grocery store or a natural-food store. Check the list of ingredients, and if they include only milk and bacterial culture, you've found what you need. Generally speaking, a yogurt labeled as "Balkan style" or "natural yogurt" is suitable.

Coagulants

Coagulants are enzymes that cut a certain bond in the protein molecule in milk. This cut bond allows the protein to stick together and form a curd. There is much more of a description to the curd forming-process than this—a topic for another book. There are four different sources of the enzymes that coagulate milk. The traditional animal rennet comes from part of the stomach of a ruminant, typically from a calf, but lamb and kid rennet can also be found. Microbial enzyme is usually produced by a certain type of mold. There are bacteria and yeasts that have been genetically engineered to produce a coagulant enzyme. Coagulant enzyme can also be extracted from a specific type of plant. All forms of coagulant must be mixed with water before adding to your milk. The coagulant you find will most likely be a tablet. There are also liquid forms available. It is important that non-chlorinated water be used. Chlorine will deactivate the enzyme and no curd will be formed.

Salty feta cheese pairs deliciously with bell peppers, cucumber and tomatoes.

- **Coagulant.** This will most likely be rennet or a microbial enzyme.
- **Salt to taste.**

Materials
- Stainless steel double boiler
- Long-handled stainless steel spoon
- Thermometer
- Teaspoons with metric measure indications
- Long, straight spoon handle or knife that you can use to cut curd
- Colander lined with cheesecloth over bucket to catch the whey

Method
1. Record date, the type of cheese and any relevant details about the milk, such as its age.
2. Wash hands and sanitize all equipment.
3. Warm water in the double boiler over medium to 30°C (85°F), then turn burner to low.
4. Add milk to milk pot of double boiler and place over the water jacket.
5. Heat milk to 25°C to 30°C (75°F to 85°F). Turn heat off and remove pot from heat. Add the live yogurt culture. (If you want to change the batch size, a general rule is 15 mL: (one tablespoon) of yogurt per liter/quart of milk.) Stir well for five minutes.
6. Cover the pot and let the milk rest at this temperature for about 45 minutes to one hour. The

bacteria from the added starter culture (yogurt) will multiply and acidify the milk during this time. Take this time to measure out and stir your coagulant into water. It should rest for about 20 minutes before adding the coagulant to the milk.

7. If you're making cheese with only four liters/quarts of milk and the tablet of coagulant you have is good for 50 or 100 liters/quarts, you can make it work. First, if you're able, break the tablet in half or into quarters. Note the volume of milk this piece of tablet is good for. If the tablet was originally good for 100 quarts/liters, the half-tablet will be good for 50 and the quarter-tablet for 25. Record this information.

8. Add the coagulant to 125 mL (half a cup) of water. (I use this quantity for easy calculation and measuring.) A small squeeze bottle is helpful.

9. You now know that you have 125 mL (half a cup) of coagulant that will coagulate 25 quarts/liters of milk.

10. Use the quantity equation on page 143 to find the amount of mixed coagulant you need for the amount of milk.

11. Store the unused portion in the fridge for the next time you make cheese. The coagulant mixture should last for around six weeks if kept in a dark, cold location. Do not freeze. Be sure to record the concentration on the bottle and in your journal.

12. Remove the lid from the pot of milk. The milk should still look like milk at this point. Add the coagulant according to the manufacturer's directions. Stir for one minute. Replace lid on the pot and let rest for an additional 45 minutes to one hour.

13. When you lift the lid the milk will appear as a solid white, opaque mass. Run the back of a finger across the top of the curd. To see if the curd is ready to cut, check for a "clean break." To do this, insert a finger straight into the curd just past the second knuckle and then gently bend the finger up. As you do this the curd should "break," leaving a straight, clean edge where your finger has broken through the curd. You can also test the curd by making a small slice in the curd with a knife. If the curd is ready, there will be a distinctive slice in the curd that fills with greenish-yellowish whey. If your curd has not reached a clean break, let it rest for another 10 to 15 minutes and check again.

14. Using the handle of a large spoon

When a clean break is observed, it is time to cut curd. See Step 13.

Carefully cutting feta curd into 1" cubes. See Step 14.

or a knife with a long, flat blade, cut the curd into 1-inch cubes. Do this by first slicing vertically from top to bottom of the vat. Make a grid pattern, being sure to cut through the curd right to the bottom of the pot. It is a little more difficult to make the horizontal cuts. You can use a flat spoon on an angle to carefully slice the curds. Don't worry if your cubes aren't perfect. It's better not to break the curd too much and have larger pieces. You can cut them as you stir.

15. Once the curd has been cut, let it rest for five minutes. After the five-minute rest, gently stir the curd with a large spoon. You must stir the curd in a special way so you don't damage the slices. Place the spoon vertically into the curd to the bottom of the pot, then gently use the spoon as a scoop and pull the curd at the bottom up to the top of the pot. As you pull the curd up, use the side of the spoon to cut any oversized pieces. You're aiming for pieces of curd of uniform size.

16. Stir the curds this way for about 20 minutes. You can take short five-minute breaks. Over the stirring time, the curd will lose its shine, become smaller and firmer, and seem drier. When the curd is ready, it looks a bit like cooked scrambled eggs.

17. When you're happy with the look of the curd, you can let the curd settle to the bottom of the pot. This should take about five minutes minimum. Scoop off about 50 percent of the whey. Pour the

Feta curd after the removal of whey. See Step 17.

Feta curd in a colander. Holes in this colander are small enough that cheesecloth is not required.

remaining curds and whey into the cheesecloth-lined colander. Push the curd into the center of the colander and fold the cheesecloth over the top. Let the cheese rest overnight in a dry and draft-free place.

18. By the next morning you should have a block of cheese. Cut the cheese into blocks that can be packed in containers and covered with brine. If you've made the cheese in the colander and it has a rounded shape, cut a large cube out of the center. Depending on the size of the leftover bits, you can leave them whole or crumble them up. Play with cheese shapes by using different cheese molds.
19. Dry salt the cheese by sprinkling the cheese with salt. The goal is to lightly cover all surfaces of the cheese. Rub briefly until the salt is evenly coating the surface of the cheese. Do not over rub. The goal is for an even distribution of salt on the cheese. If you're weighing your salt and cheese, a good place to start is 5 percent by weight.
20. Let the cheese rest again for 12 to 18 hours in a warm, draft-free environment.
21. Pack the cheese in a cheese brine (see below) and let it age for at least 24 hours. (Up to several months is fine.) You can put the cheese in plastic pails, glass jars or other containers. The important thing is to have the cheese covered in brine and be able to store the cheese in a cool location, below 10°C (50°F).

Feta curd removed from the mold after resting overnight.

To make brine, mix well:
- 1 liter/quart of water
- 80 grams (about 1/3 cup) of salt
- 2 tablespoons vinegar

Pour mixture over your feta cheese.

The feta cheese should last and be just as perfect the day you take it out of the brine as the day it goes in. A few problems may arise. If your cheese becomes very soft, the brine is not acidic enough. The lack of balance of pH between the cheese and the brine causes calcium to leach out of your cheese. Calcium is important in curd structure. The decrease in calcium in your cheese weakens the curd. The result is a soft, mushy mess.

This problem can be corrected by adding more vinegar to the brine.

Mold may grow on the surface of your brine, but it should not affect your cheese. Simply scoop the mold off the top of the brine and monitor the cheese. You may wish to change the brine.

The cheese in the brine may develop a slimy feel. This is most likely caused by yeast. If the cheese smells like bread dough, yeast is definitely present. Your cheese may be salvageable. Discard the brine and replace with fresh brine. To avoid this problem in the future, use improved hygiene, cleaning and sanitizing practices while making your cheese.

CHAPTER 9
Soap Making

IN MY OPINION, handmade soaps are the most luxurious, kind-to-the-skin, non-irritating, moisturizing product you can use to clean your face and your skin. Young skin, old skin, baby skin, dry skin, oily skin, skin with rosacea, skin with eczema, skin with acne, and allergic skin can all benefit from the use of handmade soaps, especially those made with the nourishing properties of goat milk.

Handmade soaps are simple. No detergents, no fragrances (unless you want them, but handmade soaps smell wonderful without any help), no fillers, no preservatives, no unknown chemicals. All cold-process handmade soaps are fantastic to use, but the ones with goat milk are the best. It is said that Cleopatra's beautiful skin was the result of her bathing in milk. When my oldest daughter was a baby, I added milk to her bath to help her eczema. As long as I added the goat milk, Clara's skin stayed baby-soft.

In doing my research on goat milk soaps, it amazed me how many soap makers don't make real goat milk soap. Many recipes make soap with water and then add powdered goat milk, or the soap is made from a mixture of goat milk and water. In my mind, what's the point? If you're going through all the effort of making soap from goat milk and calling it "goat milk soap," it should be made with as much goat milk as possible. No water!

Before we go any further, I will explain the elements of soap and how they come together to create a nourishing goat-milk bar.

Elements and Process

Fats

Fats are edible but oily; they can be liquid or solid, hard or soft. They are completely safe to handle until they are heated to a high enough temperature, at which oil can burn skin. Each type of fat lends different qualities to soap. See Table 2 on page 170 for a list of fats used in soap making and the qualities given to soap.

Sodium hydroxide

Sodium hydroxide (NaOH), or lye as it is commonly called, is a strong chemical that reacts readily with fats. For this reason, lye is commonly used as a grease cleaner. Lye is highly caustic. There are some precautions you must take when handling and working with it. Many people don't use cold-process soaps because of the fear that caustic lye is in the soap. There's no need for concern, however, since no lye is left in a bar of soap. It's completely used up during the chemical reaction phase of soap making.

    Lye can be difficult to find. Lye should not be shipped because of its volatile and dangerous nature. It's very important to get a 98 percent or higher pure form of sodium hydroxide. Lye is sold by soap-making supply companies, hardware stores and chemical and cleaning companies. It is often found with drain cleaners or other strong cleaners. *Do not purchase products such as Draino for soap making.* While these products do contain sodium hydroxide, they also contain many other chemicals that you don't want in your soap. Remember to look for the percent purity of the sodium hydroxide on the product. If there is no indication of purity, don't buy that product.

Glycerin

Glycerin, when not combined in soap, is a syrupy, clear liquid. It's an emollient, giving softening and soothing properties to the skin. During commercial soap production, the glycerin is most often removed, leaving the bar of soap without this skin-nourishing ingredient. The glycerin present in soap is great for your skin. By making your own soap, all the goodness of glycerin remains intact.

Process

The chemical reaction that creates soap is called saponification. A solution of sodium hydroxide is mixed with fats and liquid until well combined. The resulting thick liquid is poured into molds and left to rest for about 24 hours. The bars are then removed from the mold and allowed to cure for four to six weeks. The result is a beautiful, moisturizing bar of soap that is kind to the skin.

Of course, during soap making we cannot see molecules interacting. We can see only the resulting mixture. It's the visible change in the mixture that must be observed closely. At the beginning, the soap-making pot contains nothing but liquid fats. As the sodium hydroxide solution is added to the oils, all that's visible in the pot is a mass of separated liquids—they're separated because fats and water don't mix. It's the job of a soap maker to mix the solutions of fats and sodium hydroxide just right so that the soap-making reaction occurs with desirable results.

For the molecules to react as we want them to, we must provide perfect conditions and a little encouragement. This is relatively easy to achieve with patience and attention to detail. The first critical step is to calculate the correct amount of sodium hydroxide needed to react with our fats. Too little sodium hydroxide results in too much leftover fat in the soap. The result will be an oily product that turns rancid. Too much sodium hydroxide results in a bar of soap with leftover lye. This creates a bar of soap with a high pH that's caustic to the skin. Both are undesirable and unusable results. Below I'll outline ways to get the ratio of elements just right.

Formulating Your Soap

This is the fun part. You get to create a bar of soap that suits you. There are a variety of fat sources you can use. Each one comes with its own personality, lending qualities to the bar of soap such as hardness or softness and longevity, as well as cleaning and moisturizing properties. Coconut oil, for example, creates great lather in soap. This lather makes it excellent for cleaning and is a must in shampoos to remove the oil buildup in hair. However, this same lather and cleaning ability can also be drying to the skin.

Lather is a wonderful quality in soap—it's what makes the soap spread around and gather dirt, and it makes us feel good. Lather is made up of bubbles. Small bubbles are great, but large ones are drying to the skin. Copious amounts of bubbles are usually a sure sign of the presence of a detergent. Detergents are drying to the skin. Lots of lather made up of tiny bubbles is the best.

Table 2 is a good, general guideline. Once you decide on what oils you think you would like to use, you can

Table 2: Fats and the Properties They Lend

Fat	Bar soap property	Skin properties	Lather	Amount to incorporate
Coconut oil	Hard	Drying, good cleaning	Excellent, Big bubbles	Less than 30%
Olive oil (not virgin or extra virgin)	Hard and brittle	Very moisturizing, softens, mild cleansing	Thin, creamy, small bubbles	Up to 100%
Canola oil	Medium	Moisturizing	Dense, creamy	Up to 50%
Palm oil	Medium	Cleans well, mild	Long lasting	20–30%
Lard	Soft	Does not work well in cold water	Poor	Up to 70%
Tallow	Very hard	Poor cleansing	Poor	Up to 70%
Safflower oil	Medium	Moisturizing	Good	Up to 60%
Soybean oil (vegetable shortening/oil)	Medium	Makes a good base	Low lather, but mild, stable	Up to 50%
Sunflower oil	Soft	Moisturizing	Good	Up to 20%
Shea butter	Medium	Soothing	Creamy lather, small bubbles	Up to 20%
Cocoa butter	Brittle and hard	Softening to the skin	Creamy, small Bubbles	Up to 15%
Sweet almond oil	Medium	Conditioning	Good	Up to 10%
Corn oil	Soft	Balancing	Good	Up to 20%

always further research them online. The percent values are just a guide. You can certainly go outside these parameters, depending on the type of soap you would like to use. It's probably wise to start with a relatively simple recipe (such as the one I provide) and, after you're familiar with the soap-making process, you can start experimenting. For instance, if you find a non-goat milk soap recipe you like, you can substitute goat milk for the water listed in the recipe. You can take out or include any additives that you like. The three important factors to watch for are:

1. Which ingredients are the fats?
2. How much liquid?
3. How much lye?

As long as you're not changing the amounts of these three key factors, you can add and remove other ingredients to suit your needs. Also, note that soap recipes should be written out by weight. I would not attempt a recipe that's written by volume. Find another one.

Personally, I like creating my own recipes—and I encourage my students to be creative. You can work out your recipes using saponification calculations. This is fine if you're very good at math and know how to do the calculations. I make too many mistakes and therefore like to use what is called a lye calculator (or saponification calculator). A number of these calculators are available online.

My favorite one is the Majestic Mountain Sage (MMS) Lye Calculator. It has been around since at least 2003, when I started making soap. (It is available at thesage.com/calcs/LyeCalc.html. It's provided free of charge but is copyrighted and can't be reproduced.)

The lye calculator is simple. First, you need to have a list of the types of oils—and an idea of the percentage of each oil—you'd like to use. Then enter the figures and use the calculator.

For example, you may want to create a soap containing coconut oil and sunflower oil as a base, with a touch of cocoa butter for softness. You'll want your numbers to add up to 100, to represent 100 percent. Most likely, you'll want to keep the coconut oil relatively low at 20 percent so the bar lathers nicely without causing dryness. Cocoa butter can be up to 15 percent, so let's keep that value. We're up to 35 percent. That leaves us with 65 percent sunflower oil. To sum up:

Coconut oil, 20 percent
Sunflower oil, 65 percent
Cocoa butter, 15 percent
Total, 100 percent

Plug these numbers into your lye calculator and hit the *Calculate* button. You'll note the recipe calculated for a very small batch of 100 grams (or 3.5 ounces, depending on your unit of measure). This is enough for about two bars of soap. The weight is only for the oils. (Some of the weight of your bar will also be the liquid—in this case, goat milk.)

Since we calculated for 100 percent,

we end up with 100 g total. On the MMS calculator, after you hit calculate and get your recipe, there is a resize batch option at the bottom of the screen. A 200g bar is a nice size. If you want 20 bars of soap, resize for 200g × 20 = 4,000.

The lye calculator you use should give you a range of liquid that you need. If one calculator doesn't give you a range, find another. Lots of models are available.

Since you're making goat-milk soap and the benefits of the goat milk are desired in your finished soap, use the maximum amount of liquid recommended for your milk quantity. Don't worry about the fats present in the milk changing your recipe. There's wiggle room here. If you were to make an extremely large batch of soap—a batch bigger than you could ever make in your kitchen—then you'd worry about the milk fat. Select somewhere mid-range for your lye amount.

I like a bar of soap that's between 150 and 200 grams (5 and 7 ounces). This weight of a bar is nice to hold and will give you quite a few washes before it disappears. The shape of your soap also matters, so pick a mold that suits your likes and needs. You can also use a large mold and cut the newly cured soap into slabs or chunks of desired sizes. Use your imagination.

To size your batch, take into account the weight of all your oils and milk. You'll have to play with the numbers a bit because you calculate your recipe according to the percentages of fats and oils you'd like to use. Once you plug your values into the lye calculator, it will give you the amount of liquid needed as well as the weight of lye and oils. This liquid amount will factor into your soap weight and volume.

Remember, when formulating your own soap recipe, work in percentages first.

Setting Up for Success

Once you know what ingredients you're going to use and how much of each, the next critical step is to measure all ingredients carefully. A scale is a must. Measurements by volume are *not* accurate enough. Soap-making ingredients are too costly to ruin a batch because of inaccurate measurements. Invest the time in accurate measuring, and invest in a scale. You need a scale that will measure in grams (or ounces) and with a maximum weight large enough to weigh all your ingredients. Somewhere between

Soap Making

The items required for making soap.

3 and 6 kg (six and 12 pounds) should be adequate for most home soap makers.

Once the ingredient list has been calculated and ingredients have been weighed, it's critical that the temperatures of fats and lye solution are correct. A good thermometer is also a necessity. Temperatures of all ingredients should be as close as possible to one another at the time of mixing.

Mixing of the ingredients is another critical step. If the soap ingredients are not mixed consistently, they will separate and the soap will not turn out successfully. The best tool to use to mix soap ingredients together is an immersion blender—the same kind used to blend soups. It's worth investing in a good one. The motor will burn out if it's not sufficiently powerful. I've gone through several of these blenders and discovered that one with a 200-watt motor is powerful enough to mix soap.

The last critical step is curing the soap in the right environment. For milk soaps, a draft-free area that's at room temperature works the best. A low relative humidity is also beneficial. Unless you have air-conditioning, the hottest, most humid days of summer are not the best for curing soap.

Soap-making tools

You also need the proper equipment. No substitutions! For example, don't use a wooden spoon in place of a stainless steel one.

- 8 to 10 liter/quart stainless steel stock pot (must be big enough to hold all soap ingredients, with extra room—at least an additional 50 percent—for mixing) for heating the oils and mixing all soap ingredients
- Large stainless steel spoon with long handle for scooping ingredients and/or stirring oils
- Large stainless steel serving-size spoon (one size up from a soup spoon) for stirring lye solution
- Stainless steel soup spoon for moving small quantities of soap mixture
- Scale (one that measures in ounces or grams, up to a minimum of 3 kg/7 pounds
- Digital thermometer with stainless steel probe and/or a stainless steel dairy thermometer with clip. The dairy thermometer comes with an identifying green zone that indicates the required temperature for soap making.
- A variety of plastic containers for measuring oils (2 to 4 liter/quart ice cream containers work well, depending on your soap batch size)
- 500 mL (2-cup) container for measuring lye—should be easy to hold

- Large 2L (8-cup) glass measuring cup with pouring spout (stainless steel works, but I like the visibility with glass)
- 1 or 2 rubber spatulas for scraping oils or soaps
- Small plastic dish (half a cup) for fragrance (if using)
- Ice cubes

You will also need:
- A stainless sink to work in, or a large basin (it doesn't matter what it's made of)
- A big sink or area to wash all soap-making equipment—remember, the fresh soap will be caustic
- A good-sized work surface. Find an area large enough to hold all equipment and still leave room to work.
- A source of electricity
- A stove element or hot plate
- A plastic cloth to protect your work surface (make sure the cloth doesn't hang down. You don't want to catch the cloth and pull all your soap-making supplies to the floor.)
- Tight-fitting old clothing, or clothing you don't care about. No long, floppy sleeves, please.
- Hair tied back
- Safety glasses
- Vinegar on hand to neutralize lye or raw soap
- A room-temperature draft-free location to place your soap for aging
- A clear floor space with no steps to avoid the chance of a fall while moving hot oils, caustic lye or fresh soap
- Some type of rack for curing soap
- A pen and notebook for record keeping. Permanent markers are also useful for labeling items.
- Soap mold

Remember, choose your location carefully. Soap needs to cure for four to six weeks. You don't want to keep moving your soaps around because you need the space they are curing in.

About your soap mold

A soap mold is the framework or container that your liquid soap will sit in for 24 hours or so until it becomes a solid. You can use a variety of items for a soap mold, depending on the desired result. There are fancy, expensive molds that create bars with patterns or in shapes—geometric, animals, cars, shoes, even goats. I find soap hard to remove from elaborate molds and also feel they take away from the simplistic beauty of a handmade object. Others love the look and variety. My favorite molds for soaps are ones that are creatively recycled. I have used paper cups, cereal boxes, loofah sponges, toilet paper tubes, plastic egg cartons and wax-lined juice cartons. Anything that will not react with the soap (such as metal) and can be pried, torn or cut away from the soap can be used. I love the wax-lined juice boxes and Dixie cups because they come "pre-lined" with waxed paper. Paper, cardboard or wood molds need to be lined with waxed paper. It's a tedious job to get all the wrinkles out of the paper, but it can most certainly be done. If you don't care so much about what the outside of the soap looks like, a few wrinkles are just fine.

When sourcing a mold, keep in mind that newly cured soap is very easy to cut. You can easily cut a slab or cylinder of soap into useable-sized pieces. I suggest a mold with a shape that has a high surface area-to-volume ratio. In other words, a large, relatively flat slab is better than a cube shaped block. Soap creates heat as it cures. The shape needs to be one that releases the heat to the air, rather than hold it inside. Excessive heat can create undesirably dry, crumby soaps with holes that look like miniature caverns.

Making Soap

Read all instructions through thoroughly at least once before starting your soap-making adventure. Being prepared and knowing exactly what you're going to do will help to avoid disappointment, frustration, wasting your ingredients and the cleanup of a big mess.

1 Give yourself at least three hours of uninterrupted time. It may not take this long, but better to allocate more time than not enough.

Make sure all the equipment is out and organized and ready to go. *Put on your safety glasses.*

2 Measure your oils and additives. (See steps for measuring on page 186.) Start with these ingredients because they don't need to stay cold (like the milk) and are not dangerous (like the sodium hydroxide).

3 Fill your sink or basin with cold water and ice. The amount of ice isn't critical—the water just needs to stay cold.

4 Place your large stainless pot containing all your oils and place it over the lowest heat setting. Your target temperature should be in the range of 40 to 43°C (105 to 110°F). But there's no need to rush to get there.

The milk/lye solution being poured into warmed oils.

5 Measure your milk into your large glass measuring cup and place into your ice bath. Make sure the water level is the same as the level of your milk. This way, your milk will stay as cold as possible and your measuring cup won't float in the sink.

Heating Oils

Oils are very good at holding onto heat. This means they take a relatively long time to cool off. You'll want to turn the heat off a few degrees shy of your target temperature. It's relatively easy to warm your oils up, but can be quite time consuming and messy to cool them down. The oils can sit and wait for your lye solution to be mixed at the right temperature.

6 Weigh out your lye carefully. Use the two-cup plastic container. Be accurate with your measurements, to the gram or ounce. This is very important. Do not pour the lye into the measuring container—the lye will bounce out of the container and end up everywhere. Use a spoon to carefully place the lye in the measuring container.

7 Very *slowly* pour the lye into the cold milk while stirring. Your milk will turn a beautiful lemony color as the milk fats react with the lye and the mixture warms up.

Milk in the ice bath is ready for lye addition.

Working with Milk

Never add milk or water to the lye. Always add the lye crystals to the milk. A violent reaction can result from the improper technique and would cause your lye/milk mixture to erupt out of your container.

The trick to making beautiful milk soaps is the *slow* addition of lye into *cold* milk and not letting the lye/goat milk mixture get too hot. As lye reacts with the fats in the milk, heat is created. Many milk soaps turn out orange because lye is added too quickly, and the milk cooks. This is why the ice water bath is also important. Temperature should almost stabilize before the next addition. But it is possible to add your lye too slowly and not reach the desired temperature of 40°C (104°F). The addition of lye to milk is a fine art—one that you will perfect with practice and patience.

8 While adding your lye to your milk, monitor the temperature of your oils. Don't stop stirring your milk solution until the portion of lye added to your milk is fully dissolved and incorporated. Failing to do this will result in orange, cooked spots in your milk. Once sodium hydroxide is dissolved and combined, it is safe to take a short break from stirring to check your oils. You don't want the lye/goat milk mixture to cool too much. If at any time your heat is rising too quickly, take a break, but only if all the lye is incorporated.

9 At your final addition of the lye, check your temperature. If you're a long way from your target, carefully remove your solution from the water bath before your final addition of lye. Hopefully, by doing this you can get your mixture up to the desired temperature of 40°C (104°F). A couple of degrees either way from this target temperature is fine—the important part is that the oil mixture and the lye mixture are within a degree or two (but not more than that) of each other.

10 Be sure both your oils and lye/milk mixture are in the desired temperature range. It's best if both solutions are at the same temperature, though being within a few degrees of each other is acceptable.

Soap Making

Adding honey to the milk/lye/oil mixture.

Blending the milk/lye/oil mixture with immersion blender. Do not raise turning blender above the level of the liquid.

11 Slowly add your milk solution to your oils. Do not stir yet. The mixture will be in two distinct layers. This is good. You don't want your soap to react too quickly. Use your rubber scraper to get all the milk solution in your oils. Any lye left behind could be the difference in good soap and bad.

12 Add your measured additives at this point, but don't add your fragrances or plant material such as oats, flower buds, herbs or powders.

13 Start to stir your soap mixture with your immersion blender turned off. You should be able to stir your mixture enough by hand to incorporate the layers. Once you have a more homogenous mixture and the mixture has slightly more body to it, you can start using your immersion blender. Use short bursts to avoid burning out the motor. Don't run the motor for longer than 10 seconds or so. Allow the motor to cool for 20–30 seconds. The amount of time you can run your immersion blender in the soap depends on it quality and size. Another tip to prevent burning out the motor is to angle the blender and allow the mixture to flow up through the blades. This is especially important as the mixture thickens. You'll feel the blender suction to the bottom of the pot if it's not at a great enough angle. If the blender is on too much of an angle, the mixture will splash out of the pot. So wear safety glasses to protect your eyes from the caustic soap solution.

14 Once the mixture stops separating, but before trace is reached (see step 15), add your fragrances. If you want to have a scent in your soap, you can use perfume fragrance oils. A number of high-quality fragrance oils are available from various soap-making suppliers. Don't use fragrances that are meant for other purposes, such as candle making.

Fragrances must be added at the same temperature as your soap mixture—otherwise your mixture may separate during the fragrance addition and it will not go back. You can warm the fragrance in a warm water bath or for a few seconds in the microwave. Try five seconds to start.

Using essential oils for fragrance in cold process soap making is difficult because they don't stand up to the extreme pH of the cold process. If you want to use essential oils, you must cure your soaps first, until they achieve a neutral pH. Soaps must then be milled (a special process) and essential oils added. Don't waste your money trying to add essential oils to your uncured soap mix. I have tried it. It will not and cannot work.

15 Blend soap mixture until trace is achieved. During soap making, your goal is to achieve trace: the point at which the two mixtures have been sufficiently blended to create soap. If trace is not achieved, the soap will separate when left in the mold to cure. To know if you have reached trace, you need to use your powers of observation. Your mixture will turn from a separated mass of oil and lye/milk solution to a beautiful homogenous soap mixture. Watch the oils and lye mix together. Less and less oil will be visible in the solution as you mix. Your mixture will turn from one with a glossy finish to one of a flat finish. Your soap solution will also have thickened and have a nice body to it. The telltale sign that you've achieved trace is a "tail" that's left if you drizzle some soap to the top of your mixture. My favorite way to check for trace is to watch the side of my immersion blender where it meets the surface of the soap solution. While you're blending the soap and the blender is running, you'll observe a small swirl where the blender shaft meets the soap mixture. Once you

Trace is achieved when small ripples can be seen on the top of the soap mixture. The trails of dripped soap mixture remain on the surface for a few seconds and a small circular pattern can be seen beside the shaft of the immersion blender when turned on.

think the soap has reached trace, let it set for 10 seconds, give it a quick stir and check for trace again. If it holds and exhibits the same properties of a "trail" or "swirl," you have achieved trace.

If your soap mixture is blended too much past trace or if you're too slow pouring soap from pot into molds, your soap may seize. This means it will form a solid mass in the pot you're working in. You want the soap to solidify in your molds—not in the pot, so you need to work diligently and without interruption during soap making. This is no time to answer the phone.

16 The type of mold you'll be using will determine how thick you blend your soap. As soon as trace is achieved, you can pour your soap into your prepared mold(s). If you're using a mold with an intricate design, or several small molds (which takes time), you'll want to pour your soap at a light or thin trace. If you're pouring into a large mold that's quick and easy to fill with no tiny nooks and valleys, I would recommend blending the soap to a bit thicker trace. This way, you'll have no doubt that you have indeed reached trace.

After the soap is safely in its mold(s), use your soup spoon to move soap

Pouring soap that has reached trace into soap molds.

around if necessary. Make sure all cavities are filled and molds have an equal amount of soap. You'll be able to tell when it's time to stop playing with your soap and let it rest. It will start to become more solid and will not incorporate as nicely.

Your soap will stay in the mold for 24 to 48 hours. Some soap molds release the soap easily, and 24 hours of cure time is sufficient to achieve a hard enough bar of soap that can easily be removed from the mold. Some molds require a harder bar for the soap to be removed. The fancy molds in the shape of goats, for example, are quite difficult to remove and the soap needs to be fairly solid to be removed without damaging the shape. One way to speed up this process (especially if you want to use your mold again) is to put the mold containing the soap into the freezer. Usually 20 minutes to an hour will do it. The only drawback is that you'll slow the curing process slightly. Chemical reactions generally happen more quickly with heat. Too much heat and your soap may cure too quickly, causing undesirable results. If the environment is too cold, the reaction will take a longer time.

Once your soaps are safely removed from your molds, it's best to place them "standing up" so the greatest amount of surface area is exposed to the air. (This may not be possible, depending on the shape of your soap.) Standing the soap up also allows you to cure more soap in one location.

Curing Soap

Now it's time to wait.

It takes four to six weeks for soap to cure. During this time, your bar of soap will harden but, more importantly, the saponification reaction will continue until all of the lye has been "used up." Your end-product soap will be at a pH slightly above 7.

You really don't need to do anything to your soaps. Milk soaps create a lot of heat while the chemical reaction is taking place. For this reason, do not cover your milk soaps. If you're concerned about dust gathering on them, it is best to wait a week or so before covering. By that time, most of the lye will be used up and the risk of ruining your soap from too much heat will be far less. Cold-process soaps made without milk do not create as much heat as milk soaps do. Non-milk homemade soaps are generally covered during the curing stage.

How do you know your soaps are ready? You may wish to invest in some litmus or pH paper to test your bars.

On the packaging of the pH paper you'll find a pH range chart with corresponding colors. Take your pH paper and do not touch the end that will be doing the measuring. Get a drop of water on your fingertip and moisten a small area on your bar of soap. Take the untouched measuring end of your pH paper and rub both sides in the moistened soap.

The strip should turn color. If it does not, or only changes partially, try adding a little more water. Once your strip has changed color, compare this to the chart. Note the pH. If the soap is reading 7 or a shade between 7 and 8, your soap is fine. If the soap is reading a pH of 8 or higher, your soap needs more time to cure. If after eight weeks your soap has still not achieved a pH of less than 8, lower the amount of lye in your recipe the next time you make a batch of soap. Do not make drastic changes to your recipe—small changes are best. The calculator you used to formulate your recipe should have given you a range of lye. Use this range as a guide. If the calculator didn't give you a range, I suggest using a different calculator.

If you don't wish to purchase pH paper, or you can't find any to buy, you can do an easy hand-wash test to see if your soap is ready to use. Grab a bar of soap and was your hands with it. Wash well, lather for a count of 15 seconds (or the time it takes to say your ABCs—that's what I tell my kids), then rinse. Dry and wait five minutes or so. Your hands should feel moisturized. Milk soaps moisturize the skin. If your hands feel at all dry, your soap has not reached the desired pH. When I wash with milk soaps, my hands don't feel that I need to apply moisturizer. This is the beauty of handmade milk soaps. If your hands don't feel nicer than before you washed, cure your soaps for another week and try again.

Store your soaps in a room-temperature dry location. There's no need to worry about your soap going bad—it will only become harder with time. Be creative with your soaps. Try different shapes, scents and packaging options. Sell your soaps at craft sales, give them to your friends or keep them for yourself.

Goat Milk Soap Recipes: Sample Ingredients

I have included these few ingredient lists to get you started if you're not yet ready to create your own recipes for soap. I've kept the batch size relatively small, but the volumes are large enough to easily work with. For these smaller-batch sizes, be sure the pot you heat your oils in, which is also the pot in which you'll mix your soap, is not too big in diameter. Otherwise, your mixture will be too shallow to mix.

Practice soap

This is an inexpensive recipe to experiment with and achieve a good soap make before investing in the more expensive oils such as coconut or olive. This bar of soap will be of medium firmness and have a nice creamy lather, but few or no bubbles. This soap will allow the moisturizing properties of the goat milk to shine through.

- Canola oil (600 grams/20 ounces)
- Soy oil (vegetable oil) (600 grams/20 ounces)
- Sunflower oil (300 grams/10 ounces)
- Lye (190 grams/6.5 ounces)
- Whole goat milk (560 grams/20 ounces)

Cocoa goat soap

This soap uses olive oil as a base. 100 percent olive oil soaps are beautiful for the skin, but soaps made from a high

olive oil content are very hard. This recipe is mainly olive oil, with added coconut oil for lather and cocoa butter for softness.
- Cocoa butter (225 grams/8 ounces)
- Coconut oil (300 grams/10 ounces)
- Olive oil (975 grams/34 ounces)
- Lye (202 grams/7 ounces)
- Whole goat milk (560 grams/20 ounces)

Shampoo bar

This bar recipe has a high coconut oil content to create lather and cut through dirt and oils. Sunflower and corn oil add softness to the bar, and sweet almond oil gives nourishment. I like to use this bar as a substitute for shampoo.
- Coconut oil (750 grams/26 ounces)
- Sunflower oil (300 grams/10 ounces)
- Sweet almond oil (150 grams/5 ounces)
- Corn oil (300 grams/10 ounces)
- Lye (223 grams/8 ounces)
- Whole goat milk (560 grams/20 ounces)

Steps for Measuring

1. Turn the scale on and tare (zero) it.

2. Place the container that the oil or other ingredient will be weighed in on the scale. Note the weight of the container in your notebook—it will come in handy if you make a mistake during weighing. You can use this number to calculate the weight of your oils if the scale is not tared properly. Simply take all items off the scale, then measure the total weight of the container plus oil. Subtract the weight of the empty container. The number left will be the weight of your ingredient.

3. Press the scale's "tare" button. The scale should now read zero with the empty container on it.

4. Slowly pour or spoon your ingredient into the container that's on the scale until the desired quantity is reached. It's best to take the time to measure exactly. If your recipe calls for 85g (3 ounces) of olive oil, measure out the *precise amount* – not 84g or 86g, but 85g exactly.

5. Empty your measured oil into the large stainless pot in which you'll be heating the oils.

6. It's best to weigh each oil separately. That way, if you overpour, you can easily correct your mistake by putting excess oil back into the original container. Either use separate containers for each oil or use your rubber scraper and be sure to remove all oil from your measuring container before moving on to the next one.

7. Weighing ingredients can easily take the bulk of your soap-making time. For this reason, I like to measure out ingredients for multiple batches at once. I use large plastic ice cream containers with lids to combine and store the weighed oils. Please take the time to *label your containers* with exactly what's in them. You'll be grateful you took the time to do this.

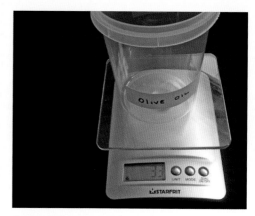

Scale measuring weight of empty container.

Scale tared to zero.

Measure 486 grams exactly.

The total amount of accurately weighted oils is in the pot ready to be heated.

Resources

Books

Mastering Artisan Cheesemaking: The Ultimate Guide for Home-Scale and Market Producers by Gianaclis Caldwell, Chelsea Green Publishing, 2012

Goat Medicine, 2nd Edition by Mary C. Smith and David M. Sherman, Wiley-Blackwell, 2009

Websites

Glengarry Cheesemaking, glengarrycheesemaking.on.ca: Canadian supplier of cheese-making ingredients and equipment for the home cheese maker.

Majestic Mountain Sage, thesage.com: Full-service supplier of raw materials for handcrafted soap and cosmetics. They also have a lye calculator and valuable information on soap making.

New England Cheesemaking Supply Company, cheesemaking.com: American supplier of cheese-making ingredients and equipment for the home cheese maker. I highly recommend signing up for their excellent newsletter.

Ontario Ministry of Agriculture, Food and Rural Affairs (OMAFRA), omafra.gov.on.ca: A website that features fact sheets and information about farming and milking goats, making cheese and the regulations that govern these activities.

River's Edge Goat Dairy, goatmilkproducts.ca: Learn more about our farm and our products, try recipes and find information on how to visit. You can also subscribe to our Facebook page (facebook.com/RiversEdgeGoatDairy).

Index

abortion, 133–134
abscesses, 121
acidosis, 116, 128, 134–135
air quality (in barn), 17, 113
Alpine (goat breed), 20
American Dairy Goat Association (ADGA), 20
ARS shears, 107
Arthur, Ontario, 8

bacteria, 87, 99–101
bedding, quality of 17
birth weight, 114
birthing positions of kids, 64–67
bloat, 136
Bloat Aid, 136
body condition score (BCS), 28, 38
Boer (goat breed), 14
bottle feeding, 76–79
bottles and nipples, 76–77
breech birth, 63–64
breeding
    basics, 46–47
    options, 48, 50
    time of, 44, 51–52
Brevibacterium linens, 158
"buck-in-a-jar", 48
buck, odor of, 47

caprine arthritis and encephalitis (CAE), 119–120
carbohydrates (in goat diet), 30
caseous lymphadenitis (CL), 121
CD-T vaccine, 123
Center for Food Security and Public Health, 130
Centers for Disease Control and Prevention, 130
cheese
    chèvre, recipe for, 151–158
    culture, starter, 152
    curd, cheese, 143
    draining, 151
    feta, recipe for, 160–165
    goat, creamy, recipe for, 148–150
    making, 139–165
    paneer, recipe for, 158–159
    ricotta cheese, recipe for, 145–147
cheesecloth, 146
chlamydia infection, 126
chlorine, 142
cleanliness, importance of, 140–142
Clostridium perfringen, 123
Clostridium tetani, 123
coagulants, 160
coccidiosis, 122
colostrum, 19, 40, 44, 70–71, 76, 81, 99–100, 120
Coxiella burnetii, 130
creep areas, 80–81
Crossbred (goat breed), 23
cud, 33, 116

dehydration, 72
diet, life stages, 40–41
digestive system, illustration of, 32
direct culture unit (DCU), 152
disbudding, 105–108
diseases
    in goats, 115–139
    symptoms of, 50
dust and mold, in hay, 39

E.coli, 101
ecthyma, contagious (sore mouth), 122–123
Eimeria, 122
electrolytes, 116
enterotoxemia, 123
estrous cycle, in goats, 114

fats
    in goat diet, 30
    in soap making, 168, 170
feed, quality of, 17
feeders, elevated, 28
feeding
    bottle, 74–75
    strategy, 38–40
    time, 73–74
fencing, 24–25
fiber (in goat diet), 32
foot rot, 123–124
forages, 39

gender, how to determine, 70–71
gestation
    calendar, 45
    in goats, 114

glycerin, in soap making, 168
goat collar, 89
Goat Medicine, 119
goat milk soap recipes, 184–185
goat
    age at purchase, 16
    crossbred, 15
    demands made on owners, 10–11
    determining gender of, 19
    diet, 34
    diseases of, 113–137
    grain as feed, 35
    illness of, 113–137
    polled, 105
    purebred, 15
    registered, 15
    size of initial heard, 25
    stand, 89
    temperature of, 115–116
    types of, 14
Guelph
    Ontario, 10
    University of, 7

hay, 34, 36
    quality of, 29
heart rate, in goats, 115–116
heat lamp, 70–71, 83
heat, length of, 114
herd, closed, 132
hoof trimming, 109–113
horns, disbudding, 105–108
housing, types of, 23–24, 83–84

illness
    in goats, 113–137
    signs of, 117–119
injections, 132
internal parasites, 137

keratoconjunctivitis (pink eye), 126–127
ketosis, pregnancy toxemia, 129–130
kidding, 57–61
    kit, 55
    multiple at birth, 66–67
Kiko (goat breed), 14
Kitchener, Ontario, 10

labor, 56–57
LaMancha (goat breed), 21
lice, 124
Listeria, 99–100
lye calculator, 171

Majestic Mountain Sage (MMS) Lye Calculator, 171
mastitis, 124–125
measuring, using scale for, 186–187
medication, 135
Merck Veterinary Manual, 119
metabolic diseases, 133–135
milk
    freeze, 95
    production cycle goat's, 96
    how to stop, 52–53
    quota, 8
    replacer, 19, 75
    working with, 178

milking, 88–94
    equipment, 87–88
    routine, 96
    stand, 88
minerals (in goat diet), 31, 36
Mycoplasma conjunctivae, 126

Nigerian Dwarf (goat breed), 21
nipples and bottles, 76
Nubian (goat breed), 14, 21
nutrients, 30–33

Oberhasli (goat breed), 22
oils, heating, 177
Orangeville, Ontario, 10
orf, *see* ecthyma
ovulation, 43–46

panting, 116
parasites, internal, 137
pasteurization, 100–103
pasteurized milk (vs. raw), 99
Pepto-Bismol, 135–136
pink eye (keratoconjunctivitis), 126–127
placenta, delivery of, 56–57
polioencephalomalacia (PEM), 128
polled goat, 105
postpartum care, 68
pregnancy toxemia (ketosis), 129–130
proteins (in goat diet), 30
Pseudomonas sp., 100, 158
puberty, age of, 114

Q-fever, 130–133
    in humans, 131
quantities, calculating, 143

rabies, 128
ration, starter, 85
raw milk (vs. pasteurized), 99
respiration, rate of in goats, 115–116
rumen, 84, 134
    bacteria, 33–37
    movement, 116–117

Saanen (goat breed), 22
Sable (goat breed), 23
salt, 148
    block, 37
sanitizing equipment, 142–143
saponification calculator, *see* lye
    calculator
scale, digital, how to use, 186–187
sick pen, 118–119
Smith, Mary C., 119
soap
    formulation, 169–172
    making, 167–185
    preparation for, 172–175
    tools for, 174–175
    molds, 175–176
    curing, 183–184
sodium bicarbonate, 136
sodium hydroxide, in soap making, 168
sore mouth (ecthyma, contagious), 122–123
South African (goat breed), 14
Special Formula, 127

stainless steel, 87, 140
standing heat, buck in, 44
starter culture, 152
straw, 34

teeth, goat, grinding of, 118, 134
temperature, of goats, 115–116
thiamine deficiency, 128
Toggenburg (goat breed), 23
total digestible nutrients (TDN), 30
trimming, hoof, 109–113

umbilical cord, 68–69
utter
    wash, 90
    wipes, 90

vaccination, 21
Vermont Institute for Artisan Cheese, 10
veterinarian, questions for, 137
Virkon, 119
vitamins (in goat diet), 20, 31

water (in goat diet), 30, 34
water trough, 84
weaning, 84–85
weight, at birth, 114
whey, uses for, 159

zoonotic, 119